AF777340

Concrete Principles

Second Edition

Workbook

AMERICAN TECHNICAL PUBLISHERS, INC.
HOMEWOOD, ILLINOIS 60430-4600

ATP Staff

The authors and publisher are grateful to the following companies and organizations for providing information, photographs, and technical assistance.

Portland Cement Association
Stanley Tools

2 3 4 5 6 7 8 9 – 09 – 9 8

Printed in the United States of America

ISBN 978-0-8269-0513-0

This book is printed on recycled paper.

Table of Contents

Introduction

Concrete Principles Workbook is designed to reinforce information presented in *Concrete Principles*. The textbook is used as a reference to complete learning activities in the workbook. Each chapter in the workbook covers information in the corresponding chapter in the textbook.

Answering Chapter Questions

The question types used in the workbook include true-false, multiple choice, completion, short answer, and matching. For true-false questions, circle the T if the statement is true, or circle F is the statement is false. For multiple choice questions, place the letter of the correct answer in the answer blank next to the question. For completion questions, write the correct term(s) in the answer blank next to the question. For short answer questions, write the correct answer in the area provided. For matching questions, place the letter of the correct corresponding answer in the answer blank next to the question.

Activities are also included in the workbook. The activities are related to the chapter in the textbook, and provide additional experiences for the reader, including math applications.

Related Information

Information presented in *Concrete Principles* and *Concrete Principles Workbook* addresses common concrete construction topics. Additional information related to concrete construction is available in other American Tech products. To obtain information about these products, visit the American Tech website at www.go2atp.com.

The Publisher

Concrete Principles

Chapter 1

Concrete

Name ______________________________ Date ______________

True-False

T F **1.** Aggregate acts as a binding agent in concrete when mixed with water.

T F **2.** Ancient Greeks were the first to use a form of concrete in 27 B.C.

T F **3.** Approximately 15,000,000 tons of portland cement are used per year for residential concrete construction applications.

T F **4.** Cement makes up the largest volume of material used in concrete.

T F **5.** Crushed ice or liquid nitrogen is a common substitute for a portion of the mix water when placing concrete in hot weather.

T F **6.** Heat rate is the rate at which heat in concrete is generated.

T F **7.** Entrained air is undesirable air voids that occur in all concrete mixtures.

T F **8.** The color in a concrete roof tile is mixed throughout the entire tile, providing a lasting finish.

T F **9.** Stucco is an exterior finish material consisting of portland cement, lime, sand, and water.

T F **10.** Water is the smallest percentage of ingredients in a concrete mixture.

T F **11.** The dry process is the most common method used to manufacture cement.

T F **12.** Calcium hydroxide is the most common accelerator admixture.

T F **13.** Green concrete is concrete that has been placed but has not yet reached full strength.

T F **14.** Welded wire reinforcement is heavy-gauge wire joined in a grid and used to reinforce and increase the tensile strength of concrete.

T F **15.** Glass fiber-reinforced concrete (GFRC) provides significant strength reinforcement.

T F **16.** Workability is the ability for plastic concrete to flow against its internal resistance.

T F **17.** A cast-in-place concrete component is a concrete component cast in its final position using moving forms.

T F **18.** The Portland Cement Association was founded in 1916 by S.T. Fowler.

T F **19.** Portland cement is classified by the EPA into seven types.

T F **20.** Water containing a small amount of sugar or citrates may be suitable for drinking but not for mixing concrete.

T F **21.** Seawater used as mix water leads to concrete with a high early strength, but a low long term strength with an overall 15% strength loss.

T F **22.** In general, a lower water-cement ratio results in a denser and stiffer cement paste.

T F **23.** Heavyweight concrete can weigh as much as 8100 lb/cu yd.

T F **24.** A colored chemical stain penetrates and reacts with the concrete below the surface to produce a permanent non-fading colored surface.

T F **25.** The aspect ratio of fibers used for fiber-reinforced concrete is the length of the fiber divided by its diameter.

T F **26.** Steel fibers help to prevent shrinkage cracks.

T F **27.** A continuous cast concrete component is a concrete component cast in final position using moving forms.

T F **28.** A No. 16 sieve is used to separate coarse and fine aggregate.

T F **29.** Water in a concrete mixture does not evaporate; it is broken down and chemically bonds with calcium.

T F **30.** Admixtures are added to concrete that has been placed in forms.

T F **31.** A roller-compacted concrete mixture normally uses 40% less water and 30% less cement than a standard concrete mixture.

T F **32.** Water-reducing admixtures are produced from lignosulfonates, which are waste products generated from manufacturing wood pulp.

T F **33.** A large quantity (0.2% to 1%) of sugar prevents cement from setting.

T F **34.** Fly ash concrete is more expensive than standard concrete and has improved placability and finishability.

T F **35.** Calcium chloride causes rebar to rust.

T F **36.** Natural color pigments are more desirable than synthetically-produced pigments because of their high degree of purity, color uniformity, and resistance to fading.

Multiple Choice

______ **1.** ___ is a mixture of cement, aggregate, and water.
- A. Mortar
- B. Concrete
- C. Pozzolan
- D. Grout

______ **2.** A solution with a pH less than 7 is ___.
- A. neutral
- B. acidic
- C. alkaline
- D. none of the above

______ **3.** ___ is a mixture of cement and water without aggregate that acts as a binding agent in concrete.
- A. Slurry
- B. Tabby
- C. Chert
- D. Paste

______ **4.** A concrete mixture that uses aggregate such as vermiculite, perlite, or pumice to reduce the overall weight of the concrete is ___ concrete.
- A. heavyweight
- B. roller-compacted
- C. lightweight
- D. air-entrained

______ **5.** A ___ is a filtering device consisting of a screen with openings of specific sizes and is used to classify aggregate.
- A. sieve
- B. mill
- C. tamper
- D. grill

______ **6.** Concrete should not be allowed to freeze before it has reached a compressive strength of ___ psi.
- A. 150
- B. 300
- C. 500
- D. 1000

______ **7.** Only mixtures with a compressive strength in excess of ___ psi are considered structural concrete.
- A. 500
- B. 1000
- C. 2500
- D. 5000

_______________ **8.** A No. 6 rebar has a ___″ diameter.

A. ⅜
B. ½
C. ¾
D. 1½

_______________ **9.** ___ is a concrete or mortar mixture transported through a hose and projected at high velocity from a nozzle to a form surface in order to produce accumulated thicknesses up to 12″.

A. Roller-compacted concrete
B. Steel fiber reinforced concrete
C. Exposed concrete
D. Shotcrete

_______________ **10.** Aggregate that has absorbed the maximum amount of moisture with no excess moisture on the outside is ___ aggregate.

A. exposed
B. saturated surface dry
C. gap-graded
D. coarse-graded

_______________ **11.** A(n) ___ is a component added to concrete to produce desired characteristics.

A. admixture
B. paste
C. clinker
D. slurry

_______________ **12.** Aggregate ___ is the condition that occurs when aggregate settles because a mixture is too thin to support the aggregate.

A. Viscosity
B. Flowability
C. Workability
D. Segregation

_______________ **13.** A ___ concrete component is formed, placed, and cured to a specific strength at a location other than its final installed location.

A. precast
B. cast-in-place
C. continuous cast
D. none of the above

_______________ **14.** ___ concrete is decorative concrete in which the cement paste of the concrete is removed.

A. Exposed aggregate
B. Reinforced
C. Architectural
D. Steel fiber reinforced

_______________ **15.** The ___ is a federal agency that requires all employers to provide a safe environment for their employees.

A. Construction Specifications Institute (CSI)
B. American Society of Concrete Contractors (ASCC)
C. Occupational Safety and Health Administration (OSHA)
D. Federal Communications Commission (FCC)

_______________ **16.** Evaporation rates of mix water greater than ___ lb/sq ft of exposed concrete surface per hour must be avoided to ensure satisfactory curing and prevent plastic cracking.

A. 0.01
B. 0.1
C. 1.0
D. 10

_______________ **17.** A ___ mixture has a 0″ to 2″ slump.

A. flowing
B. wet
C. low/medium
D. stiff

_______________ **18.** Cement comprises approximately ___ of the concrete mixture.

A. 3% to 7%
B. 6% to 10%
C. 7% to 14%
D. 10% to 15%

_______________ **19.** The ___ zone of a kiln has temperature of 2200°F to 2700°F.

A. drying and preheating
B. calcining
C. sintering (burning)
D. none of the above

_______________ **20.** Maximum aggregate size should not exceed ___ of the finished concrete thickness.

A. one-eighth
B. one-fourth
C. one-half
D. three-fourths

_______________ **21.** Concrete that is not reinforced possesses high compressive strength but has ___ the tensile strength of reinforced concrete.

A. one-tenth
B. one-fourth
C. one-third
D. one-half

_______________ **22.** The ___ zone of a kiln has a temperature of 1100°F to 1650°F.

A. drying and preheating
B. calcining
C. sintering (burning)
D. none of the above

______________ **23.** Water with a pH of ___ is suitable for concrete mix water.
A. 2 to 4
B. 4 to 6
C. 6 to 8
D. 8 to 10

______________ **24.** The fiber material used for fiber-reinforced concrete is made from ___.
A. nylon
B. polyethylene
C. polyester
D. none of the above

______________ **25.** When the volume of fiber in fiber-reinforced concrete exceeds ___%, the concrete can be difficult to mix and place.
A. 2
B. 5
C. 8
D. 10

______________ **26.** In 1818, ___ discovered rock in New York that could be used to make hydraulic cement with very little processing.
A. S.T. Fowler
B. Robert Lesley
C. Joseph Aspdin
D. Canvass White

______________ **27.** The American Concrete Institute was formed in ___ to develop a standardized means for making concrete durable, usable, and safe.
A. 1898
B. 1905
C. 1926
D. 1952

______________ **28.** ___ is a compound that hydrates slowly and contributes to strength after several days.
A. Tricalcium silicate
B. Dicalcium silicate
C. Tricalcium aluminate
D. Calcium sulfate

______________ **29.** Aggregate used in the production of concrete is commonly classified as coarse or fine using a ___.
A. clinker
B. scale
C. sieve
D. none of the above

______________ **30.** Hot weather concrete placement procedures are required when the ambient temperature is ___°F or higher.
A. 65
B. 80
C. 90
D. 105

_______________ **31.** ___ is a compound that contributes to concrete strength during the first seven days after concrete placement.

A. Tricalcium silicate
B. Dicalcium silicate
C. Tricalcium aluminate
D. Calcium sulfate

_______________ **32.** In 1860, ___ obtained a patent for a reinforced concrete wall.

A. S.T. Fowler
B. Robert Lesley
C. Joseph Aspdin
D. Canvass White

_______________ **33.** The ___ zone of a kiln has a temperature of 70°F to 1650°F.

A. drying and preheating
B. calcining
C. sintering (burning)
D. none of the above

_______________ **34.** Mix water should not exceed ___°F.

A. 75
B. 100
C. 125
D. 140

_______________ **35.** ___ is a compound that is ground with cement to retard the hydration rate of the alumina and ferrite phases.

A. Tricalcium silicate
B. Dicalcium silicate
C. Tricalcium aluminate
D. Calcium sulfate

Completion

_______________ **1.** ___ is hard, granular material that is mixed with cement to provide structure and strength in concrete.

_______________ **2.** A ground and calcined mixture of limestone, shells, cement rock, silica sand, clay, shale, iron ore, gypsum, and clinker is ___.

_______________ **3.** ___ is a porous, whitish, flint-like quartz.

_______________ **4.** ___ is the deformation of a concrete member due to sustained loads.

_______________ **5.** ___ is the process of creating a close arrangement of solid particles in fresh concrete during placement by reducing the voids between the particles.

_______________ **6.** The ___ ratio is the ratio of pounds of water to pounds of cement per unit of concrete.

_______________ **7.** ___ is a chemical reaction between cement and water that bonds molecules, resulting in hardening of a concrete mixture.

______________ **8.** A(n) ___ is a shallow, conical depression in a concrete surface.

______________ **9.** The hardening of concrete by chemical action is ___.

______________ **10.** Cold weather concrete placement procedures are required when the mean daily temperature is less than ___°F.

______________ **11.** A(n) ___ is a marble-sized pellet produced by burning a raw material in a kiln.

______________ **12.** ___ concrete is a concrete mixture that has increased tensile strength due to tensile members placed in the concrete.

______________ **13.** The ___ process is a shotcrete application process in which cement and damp aggregate are mixed in a mechanical feeder.

______________ **14.** ___ are small brick-like concrete blocks placed together to cover large areas.

______________ **15.** In the early 1500s, the Spanish introduced a type of concrete called ___ that consisted of lime, sand, and aggregate of gravel, shells, or stones mixed with water.

______________ **16.** ___ is a comparison of the mass of a sample volume compared to an equal volume of water.

______________ **17.** A(n) ___ is a liquid mixture with just enough water to make the materials fluid.

______________ **18.** Bags of cement for consumer use generally contain 1 cu ft of cement and weigh ___ lb.

______________ **19.** ___ is the use of water to cover concrete during the curing process.

______________ **20.** ___ are fine surface cracks.

______________ **21.** A steel bar containing lugs that allow bars to interlock with concrete is known as ___.

______________ **22.** Measuring the flowability of concrete is accomplished using a(n) ___ with a precision tapered cone.

______________ **23.** A(n) ___ is a large oven used to heat material at a constant temperature.

______________ **24.** ___ is the measured maximum resistance of concrete to axial loading, which is expressed as a force per cross-sectional area.

Short Answer

1. Outline the history and evolution of concrete, citing the major developments in concrete technology. Describe the impact that these developments had on society.

____

____

____

____

____

Matching

Admixtures

________	**1.** Water-reducing	**A.** Helps cement paste stick to aggregate and improves the adherence of fresh concrete to hardened concrete
________	**2.** Superplasticizer	**B.** Shortens setting time and increases early strength
________	**3.** Retarder	**C.** Increases concrete strength and workability and reduces cement content and hydration heat
________	**4.** Accelerator	**D.** Delays the setting time and hardening of concrete
________	**5.** Air-entraining	**E.** Adds microscopic air bubbles to concrete
________	**6.** Bonding	**F.** Provides high initial slump without lowering concrete strength, increasing waiting time for finishing, or segregating the concrete mixture

Portland Cement Classifications

________	**1.** Type I	**A.** Finished concrete is exposed to severe sulfate action
________	**2.** Type II	**B.** Strength is required at an early age
________	**3.** Type III	**C.** There is no exposure to sulfates
________	**4.** Type IV	**D.** Heat generated during hydration must be kept to a minimum
________	**5.** Type V	**E.** Finished concrete is exposed to moderate sulfate attack

Concrete Mixture Design

________	**1.** High-early-strength	**A.** Uses magnetite or steel punchings as aggregate
________	**2.** Air-entrained	**B.** Has microscopic air bubbles in the cement paste
________	**3.** Heavyweight	**C.** Uses Type III portland cement
________	**4.** Roller-compacted	**D.** Uses processes and additives to achieve a desired form or color
________	**5.** Decorative	**E.** Compacted with a roller and contains less water than a standard mixture

Industry and Standards Organizations

______	**1.** ACI
______	**2.** PCA
______	**3.** CRSI
______	**4.** NCMA
______	**5.** CFA
______	**6.** ASCC
______	**7.** CSI

A. Association that provides technical information and products, continuing education, professional conferences, and product shows

B. Organization whose goal is to further engineering and technical education, scientific research, and the development of standards

C. Founded by concrete contractors and other groups that provide services and goods to the concrete industry

D. Nonprofit association representing concrete foundation contractors and suppliers

E. Association representing producers and fabricators of steel reinforcement

F. Association representing manufacturers of concrete masonry and of interlocking paving and segmental retaining wall systems

G. Association representing cement companies in the United States and Canada

Dry Mix Shotcrete Process

______	**1.** Cone valve
______	**2.** Nozzle
______	**3.** Air
______	**4.** Rotating feed wheel
______	**5.** Water port
______	**6.** Shotcrete spray
______	**7.** Upper feed chamber
______	**8.** Agitator
______	**9.** Material hose
______	**10.** Lower working chamber
______	**11.** Compressed air

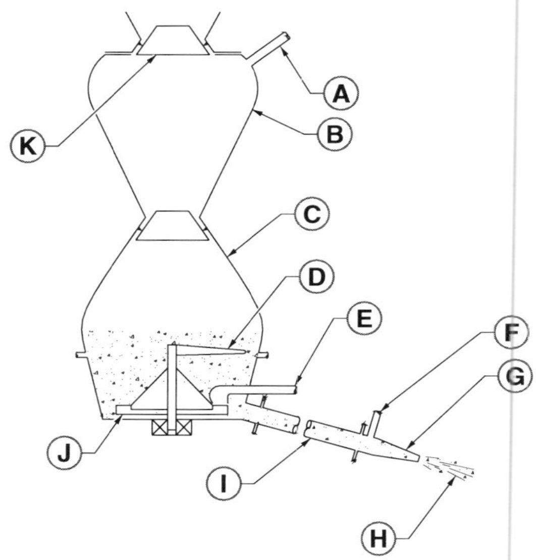

Wet Mix Shotcrete Process

____________________ **1.** Air slug

____________________ **2.** Nozzle

____________________ **3.** Air

____________________ **4.** Compressed air

____________________ **5.** Mixing paddle with rubber blade

____________________ **6.** Shotcrete spray

____________________ **7.** Material slug

____________________ **8.** Feed

____________________ **9.** Water

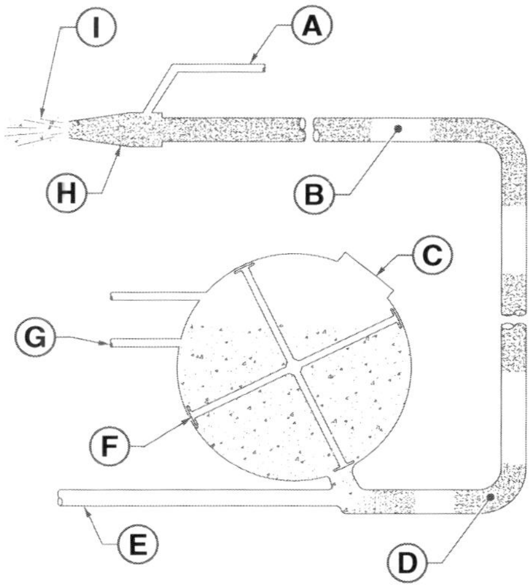

Deformed Steel Reinforcing Bars

____________________ **1.** Main rib

____________________ **2.** Steel type

____________________ **3.** Bar size

____________________ **4.** Initial of producing mill

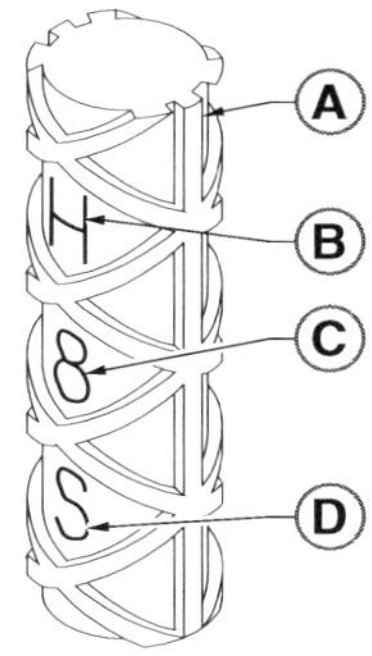

Activities

Water-Cement Ratios

The water-cement ratio is the ratio of pounds of water to pounds of cement per unit of concrete. The lower the water-cement ratio, the stronger the cement paste and concrete. More gallons of water per bag of cement yield a lower compressive strength. To determine the water-cement ratio, the weight of the water is divided by the weight of the cement. One gallon of water weighs 8.33 lb and one bag of cement weighs 94 lb. The weight of the water is calculated by multiplying the number of gallons by 8.33. A one-bag mixture using 5.5 gal. of water (45.82 lb) has a water-cement ratio of 0.49 (45.82 ÷ 94 = 0.49).

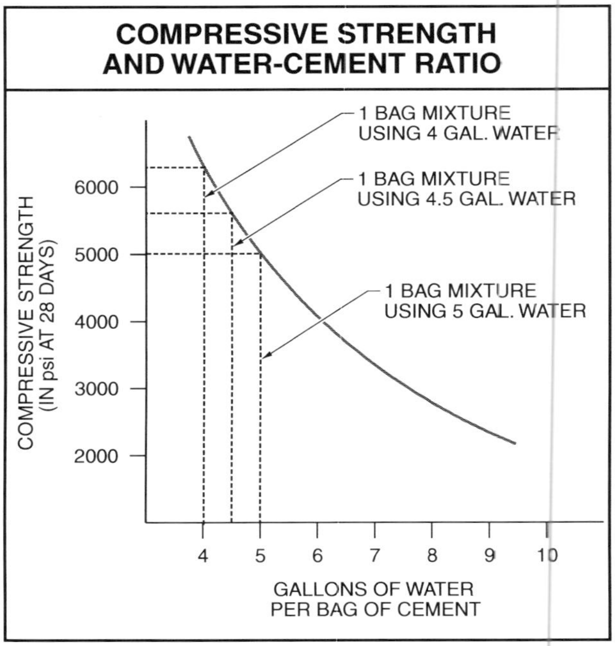

Determine the water-cement ratios for the following mixtures:

____________________ **1.** One-bag mixture using 5 gal. of water

____________________ **2.** One-bag mixture using 4 gal. of water

____________________ **3.** One-bag mixture using 4.5 gal. of water

____________________ **4.** Two-bag mixture using 10.5 gal. of water

____________________ **5.** Two-bag mixture using 8.5 gal. of water

Projects

1. Contact cement and admixture manufacturers to request literature regarding their products. Review the product literature and identify characteristics and applications for each product.
2. Research concrete construction trade associations by visiting their web sites or by requesting print literature from the associations. Identify and list each association's purpose and benefits of membership.

Concrete Principles

Chapter 2

Tools, Equipment, and Safety

Name ______________________________ **Date** ______________

True-False

T F **1.** Square-ended shovels with long handles work well for moving concrete from one location to another.

T F **2.** A check rod is used when concrete is plastic and before bleedwater appears.

T F **3.** An external vibrator is mounted to the outside surface of forms and does not contact the concrete.

T F **4.** A control joint should not be sawed until the concrete has hardened for a minimum of 24 hr.

T F **5.** The maximum intended load of a scaffold is the total of all loads including the working load, the weight of the scaffold, and any other loads that may be anticipated.

T F **6.** Safety nets should be used for workers at any level over 25′ when the workers are not otherwise protected by body belts, harnesses, or other approved fall protection.

T F **7.** Lifelines are anchored below the work area.

T F **8.** Safety nets are certified by their manufacturer to withstand a 50′ drop of a 350 lb bag of sand, 24″ in diameter.

T F **9.** A hazardous material stored in a different container than originally supplied from the manufacturer does not need to be labeled.

T F **10.** If an MSDS is not provided by a supplier, an employer must contact the manufacturer, distributor, or importer to obtain the missing MSDS.

T F **11.** An accident report is a document that details facts about an accident occurring on a job site.

T F **12.** A darby float is normally used in a standing position.

T F **13.** A safety roller is used to produce a rough and skid-resistant texture on concrete surfaces.

T F **14.** Bleedwater can be removed from the concrete surface using a floor squeegee.

T F **15.** A power screed is an engine-driven screed used for leveling the surface of a concrete slab.

T F **16.** A walk-behind power trowel has two or three rotors.

T F **17.** A pawl lock is a pivoting hook mechanism attached to the fly section(s) of an extension ladder that is used to hold the fly section(s) at the desired height.

T F **18.** Based on the three-point contact method, two feet and one hand or two hands and one foot should be in contact with the ladder rungs at all times.

T F **19.** Workers or operators using tight-fitting respirators cannot have beards or mustaches that interfere with the respirator seal to the face.

T F **20.** Metal floats should be used on non-air-entrained concrete.

T F **21.** A trowel is a hand tool with a broad, flat blade used to smooth, finish, and help compact concrete.

T F **22.** When setting up a chute, a slope with a ratio between 1:4 and 1:6 is recommended.

T F **23.** The most efficient concrete buckets are cylindrical with center discharge gates.

T F **24.** When climbing a ladder, upward progress should be caused by the push of the leg muscles and not the pull of the arm muscles.

T F **25.** Climbing the cross braces of scaffolding is an acceptable means of access to the scaffold platform.

T F **26.** When lifting an object from the ground, the back should be kept as straight as possible.

T F **27.** Bleedwater is water that rises to the surface of fresh concrete.

T F **28.** A ride-on power trowel has two or three rotors.

T F **29.** The head of an external vibrator comes into direct contact with concrete.

T F **30.** A ladder may be used for horizontal work if it is properly braced.

T F **31.** No specific background color is required for specific hazards in the NFPA Hazard Signal System.

T F **32.** Areas of the skin that have been in contact with an irritant should be cleaned using detergent.

Multiple Choice

______________ **1.** A ___ is a hand tool with a long handle and a steel grill base used for compacting fresh concrete and forcing coarse aggregate below the surface.

A. spreader
B. float
C. tamper
D. check rod

_______________ **2.** A(n) ___ is a hand tool containing a flat face with a projecting rib used to form control joints or grooves in a slab before it hardens.
A. edger
B. float
C. groover
D. trowel

_______________ **3.** A(n) ___ is a power-driven endless strap that passes over rollers, providing a moving surface on which loose materials or small articles are carried from one point to another.
A. power buggy
B. elephant trunk
C. power screed
D. belt conveyor

_______________ **4.** A ___ is the track of a ladder safety system consisting of a flexible cable or rigid rail secured to a ladder or structure.
A. carrier
B. safety sleeve
C. pawl lock
D. halyard

_______________ **5.** All scaffolds ___′ or more above ground must have guardrails, midrails, and toeboards.
A. 5
B. 7
C. 10
D. 25

_______________ **6.** When used as a free-standing unit, the height of a sectional metal-framed scaffold must not exceed ___ times its minimum base dimension.
A. two
B. four
C. six
D. ten

_______________ **7.** A ___ scaffold is supported by overhead wire ropes.
A. sectional metal-framed
B. pole
C. suspension
D. none of the above

_______________ **8.** Per OSHA, a ___ is a person who is capable of identifying existing and predictable hazards in surroundings or working conditions.
A. supervisor
B. competent person
C. training specialist
D. qualified trainer

_______________ **9.** A(n) ___ is a metal funnel-shaped container with a gate mechanism at the bottom for controlling the flow of concrete.
A. concrete pump
B. concrete bucket
C. chute
D. elephant trunk

_______________ **10.** A ___ hazard is a solid, liquid, gas, mist, dust, fume, and/or vapor that exerts toxic effects by inhalation, absorption, or ingestion.
A. flammability
B. reactivity
C. specific
D. chemical

_______________ **11.** According to the Bureau of Labor Statistics, a typical foot injury is caused by objects falling less than 4′ and having an average weight of ___ lb.
A. 10
B. 25
C. 50
D. 65

_______________ **12.** A ___ hazard is the degree of susceptibility of materials to burning based on the form or condition of the material and its surrounding environment.
A. flammability
B. reactivity
C. specific
D. chemical

_______________ **13.** In 1986, the ___ regulations covering small quantity generators of hazardous waste went into effect.
A. Occupational Safety and Health Administration
B. American Concrete Institute
C. Resource Conservation and Recovery Act
D. none of the above

_______________ **14.** A concrete ___ tool is used to place a final texture or design on the surface of concrete.
A. placing
B. finishing
C. detailing
D. none of the above

_______________ **15.** A ___ is a hand tool consisting of a rectangular piece of metal with straight edges and a concave profile.
A. tamper
B. straightedge
C. rake
D. spreader

____________________ **16.** Floats are made of ___.

A. aluminum
B. wood
C. magnesium
D. all of the above

____________________ **17.** A ___ is a magnesium or aluminum straightedge 2″ wide, 4″ high, and 8′ to 12′ long.

A. highway straightedge
B. channel float
C. bull float
D. check rod

____________________ **18.** A sign of occupational irritant contact dermatitis can include ___.

A. blisters
B. redness of the skin
C. scales
D. all of the above

____________________ **19.** A(n) ___ is a flexible tubular device used for placing concrete into deep or narrow forms.

A. concrete pump
B. concrete bucket
C. chute
D. elephant trunk

____________________ **20.** A ___ is used to establish an accurate horizontal surface of even altitude.

A. level
B. screed
C. straightedge
D. none of the above

____________________ **21.** The head of a laser transit level revolves at a maximum rate of ___ rpm.

A. 120
B. 240
C. 360
D. 480

____________________ **22.** Extension ladders are positioned on a ___ ratio.

A. 2:1
B. 4:1
C. 8:1
D. 10:1

____________________ **23.** A(n) ___ is a unit used to express the relative intensity of sound.

A. ohm
B. decibel
C. micron
D. resistor

_______________ **24.** Never stand on the top ___ steps of a stepladder.
A. two
B. three
C. four
D. five

_______________ **25.** Mesh used for bodily fall protection is normally ___.
A. 2″ × 2″
B. 6″ × 6″
C. 12″ × 12″
D. 18″ × 18″

_______________ **26.** Occupational irritant contact dermatitis should be ___.
A. treated promptly
B. soaked in warm water
C. ignored until the end of the shift
D. none of the above

_______________ **27.** A ___ is a tool used to screed (strike off) concrete to a smooth surface.
A. straightedge
B. buggy
C. safety roller
D. none of the above

_______________ **28.** An ___ number is a value indicating the flatness of a concrete surface.
A. F_L
B. F_V
C. F_M
D. F_F

_______________ **29.** The connecting line between a carrier and a safety belt must be less than ___″.
A. 3
B. 6
C. 9
D. 12

_______________ **30.** Ear protection devices are assigned a ___ rating based on the noise level reduced.
A. decibel
B. canal
C. noise reduction
D. none of the above

_______________ **31.** A ___ is a hollow magnesium or aluminum straightedge 4″ wide, 2″ high, and 8′ to 16′ long.
A. highway straightedge
B. channel float
C. bull float
D. check rod

_______________ **32.** Lightweight concrete buckets are used with concrete that has aggregate smaller than 2″ and slumps up to ___″.
A. 2
B. 3
C. 4
D. none of the above

_______________ **33.** A ___ is a gasoline-powered machine with a front-end bucket for moving concrete or other materials on a job site.
A. power screed
B. power buggy
C. belt conveyor
D. concrete pump

_______________ **34.** A(n) ___ screed is a hand-held screed powered by a small two-stroke cycle or four-stroke cycle engine.
A. oscillating
B. vibratory truss
C. vibratory wet
D. roller

_______________ **35.** The NFPA Hazard Signal System uses a ___-color diamond sign to display basic information about hazardous materials.
A. three
B. four
C. five
D. six

_______________ **36.** A ___ is a moving element with a locking mechanism that is connected between a carrier and a worker harness or body belt.
A. carrier
B. lifeline
C. fly section
D. safety sleeve

_______________ **37.** A ___ is a barrier on a scaffold that is used to keep tools and other objects from falling.
A. guardrail
B. midrail
C. toeboard
D. none of the above

_______________ **38.** Guardrails, midrails, and toeboards should be installed on all open sides and ends of scaffold platforms more than ___′ above the ground.
A. 5
B. 10
C. 15
D. 20

_______________ **39.** When pumping concrete, maximum aggregate size is less than or equal to ___ of the diameter of the pipeline.

A. one-fourth
B. one-third
C. one-half
D. three-fourths

_______________ **40.** Net lining mesh should be less than ___″ and constructed of twine equal to or greater than No. 18.

A. 1
B. 2
C. 4
D. 8

_______________ **41.** Impact-load testing consists of dropping a 400 lb bag of sand, not more than 30″ in diameter, from a height of ___″ above the highest surface at which workers are exposed to fall hazards.

A. 42
B. 50
C. 75
D. 100

_______________ **42.** Health hazards are indicated on a ___ background in the left diamond of the NFPA Hazard Signal System.

A. red
B. blue
C. yellow
D. none of the above

_______________ **43.** ___ is an inflammation caused by irritants found on a job site that come into direct contact with the skin.

A. Silicosis
B. Occupational irritant contact dermatitis
C. Emphysema
D. Congestion

_______________ **44.** A ___ is a hand tool consisting of a flat piece of metal with corrugated teeth on one edge.

A. tamper
B. straightedge
C. rake
D. spreader

_______________ **45.** A(n) ___ is used to place concrete at a remote or distant location through the use of a boom.

A. elephant trunk
B. concrete pump
C. chute
D. concrete bucket

_______________ **46.** ___ are eye protection devices that cover the entire face.
- A. Safety glasses
- B. Goggles
- C. Face shields
- D. Lenses

_______________ **47.** A(n) ___ is a power screed that consists of a frame that spans across the surface of fresh concrete.
- A. vibratory truss screed
- B. vibratory wet screed
- C. roller screed
- D. straightedge

_______________ **48.** A halyard must be a minimum of ___″ in diameter with a minimum breaking strength of 825 lb.
- A. ⅛
- B. ⅜
- C. ¾
- D. 1

_______________ **49.** When mobile scaffolds are used on concrete flooring, the floor surface must be free from pits, holes, or obstructions and must be within ___° of level.
- A. 3
- B. 5
- C. 10
- D. 15

_______________ **50.** Flammability hazards are indicated on a ___ background in the top diamond of the NFPA Hazard Signal System.
- A. red
- B. blue
- C. yellow
- D. none of the above

_______________ **51.** ___ silicosis occurs where exposure to concentrations of crystalline silica dust is highest and can cause symptoms to develop within a few weeks to four years to five years after the initial exposure.
- A. Advanced
- B. Chronic
- C. Accelerated
- D. Acute

_______________ **52.** ___ silicosis results from exposure to high concentrations of crystalline silica dust and develops five years to ten years after the initial exposure.
- A. Advanced
- B. Chronic
- C. Accelerated
- D. Acute

________________ **53.** A(n) ___ is printed material used to relay hazardous material information from the manufacturer, importer, or distributor to the employer or employee.
A. NFPA Hazard Signal System
B. Right to Know Label
C. Material Safety Data Sheet
D. none of the above

Completion

______________ **1.** A(n) ___ is any object, such as sewer pipe, water pipe, or electrical conduit, that protrudes through the top surface of a concrete slab.

______________ **2.** A(n) ___ is a metal trough used to place concrete directly into forms from a ready-mix truck.

______________ **3.** A(n) ___ is a structure consisting of two siderails joined at intervals by steps or rungs for climbing up and down.

______________ **4.** Wire, fiber, or synthetic rope used for suspension scaffolds must be capable of supporting at least ___ times the maximum intended load.

______________ **5.** A(n) ___ is a flexible line of rope, wire rope, or strap that generally has a connector at each end.

______________ **6.** A(n) ___ is a deceleration device that travels on a lifeline to automatically engage a vertical or horizontal lifeline by friction.

______________ **7.** ___ is a disease of the lungs caused by inhaling dust containing crystalline silica particles.

______________ **8.** Gloves made from ___ or neoprene provide protection from chemicals, resist puncture, and allow good dexterity.

______________ **9.** A(n) ___ is a flat or slightly rounded plate used to smooth the surface of fresh concrete before it is troweled.

______________ **10.** A concrete ___ tool is a hand tool used to control the location and grade of concrete when placed in forms.

______________ **11.** ___ is the process of leveling fresh concrete by moving a straightedge back and forth across the concrete.

______________ **12.** A(n) ___ is a temporary or movable platform and structure for workers to stand on when working at a height above the floor.

______________ **13.** A multiple-point suspension scaffold is a suspension scaffold supported by ___ or more ropes.

______________ **14.** A(n) ___ material is a material capable of posing a risk to health, safety, and property.

______________ **15.** A(n) ___ space is a space large enough and configured so a worker can physically enter and perform assigned work, has limited or restricted means for entry and exit, and is not designed for continuous worker occupancy.

______________ **16.** The tip of a single or extension ladder must be at least ___′ above the roof line or top support.

______________ **17.** The minimum dimension of the base of a sectional metal-framed scaffold when ready for rolling must be at least ___ of the height.

______________ **18.** The openings between adjacent scaffold platform planks should not be greater than ___″.

______________ **19.** A(n) ___ is an ear protection device inserted into ear canals and made of moldable rubber, foam, or plastic.

______________ **20.** A(n) ___ is made of rope or webbing and is used to catch and protect a falling worker.

______________ **21.** A(n) ___ hazard is the degree of susceptibility of materials to release energy by themselves or by exposure to certain conditions or substances.

______________ **22.** A(n) ___ is a hand tool used to produce a finished radius along the edge of a concrete slab.

______________ **23.** A finishing ___ is a detailing tool used to produce a brushed surface on concrete.

______________ **24.** A(n) ___ scaffold is a wood scaffold with one or two sides firmly resting on the floor or ground.

______________ **25.** Scaffold platform planks consist of ___″ structural planks.

______________ **26.** ___ are equipped with impact-resistant lenses, reinforced frames, and side shields.

______________ **27.** ___ are an eye protection device secured on the face with an elastic headband and may be used over prescription glasses.

______________ **28.** A(n) ___ hazard is the likelihood of a material to cause, either directly or indirectly, temporary or permanent injury or incapacitation due to an acute exposure by contact, inhalation, or ingestion.

______________ **29.** A(n) ___ is a detailing tool consisting of a molded pattern that provides a finished texture to concrete.

______________ **30.** A(n) ___ is a flat pad used by a worker to distribute weight when hand troweling fresh concrete.

______________ **31.** A(n) ___ is a heavy-duty cutter used to cut welded wire fabric reinforcement, rebar, bolts, wire rope, and rods.

______________ **32.** The hangers, U-bolts, brackets, and other hardware used for constructing a two-point or multiple-point suspension scaffold must be capable of sustaining ___ times the maximum intended load.

______________ **33.** ___ is the size of the openings between the rope or twine of a net.

Short Answer

1. A section of public sidewalk measuring 4′ wide by 12′ long must be replaced. The sidewalk is easily accessible by a ready-mix truck. List the tools and equipment necessary to place and finish the concrete for the sidewalk and describe the use of each.

____

____

____

____

2. Ladders are a common means to access the tops of wall forms and the upper floors of a multilevel structure. List five precautions that should be observed for proper safety when using a ladder.

____

____

____

____

3. Irritant contact dermatitis (ICD) is an inflammation caused by irritants found on a job site that come into direct contact with the skin. List three precautions that should be taken to prevent or reduce the occurrence of ICD.

____

____

____

____

Matching

Floats

______________	**1.** Bull	**A.** 2′ to 4′ long with 1 to 3 handles
______________	**2.** Darby	**B.** 3′ to 10′ long with a long sectional handle containing an up-and-down knuckle joint-mounted in the middle
______________	**3.** Hand	**C.** 4′ to 12′ long with a flat bottom, rounded lower edges, and radiused ends
______________	**4.** Channel	**D.** 1′ to 2′ long with a single handle gripped by one hand

Ladder Duty Ratings

______________	**1.** Type IAA	**A.**	Medium-duty, commercial, 225 lb capacity
______________	**2.** Type IA	**B.**	Extra heavy-duty, industrial, 300 lb capacity
______________	**3.** Type I	**C.**	Light-duty, household, 200 lb capacity
______________	**4.** Type II	**D.**	Heavy-duty, industrial, 250 lb capacity
______________	**5.** Type III	**E.**	Special-duty, industrial, 375 lb capacity

Ladders

______________	**1.** Fixed	**A.**	Adjustable-height ladder with a fixed bed section and sliding, lockable fly section(s)
______________	**2.** Single	**B.**	Folding ladder that stands independently of support
______________	**3.** Extension	**C.**	Ladder that is permanently attached to a structure
______________	**4.** Step	**D.**	Fixed-length ladder with only one section

Levels

______________	**1.** Carpenter's	**A.**	Telescope can be adjusted vertically and horizontally to establish straight-line references
______________	**2.** Transit	**B.**	Air bubble located inside a vial filled with liquid establishes plumb and level references
______________	**3.** Laser transit	**C.**	Laser beam and receiver establishes level and plumb references

Concrete Bucket

_______________ **1.** Rear wheels

_______________ **2.** Bucket

_______________ **3.** Operator handlebars

_______________ **4.** Front wheels

_______________ **5.** Operator platform

Vibratory Truss Screed

_______________ **1.** Frame

_______________ **2.** Finished grade level

_______________ **3.** Engine

_______________ **4.** Winch

Extension Ladder

_______________ **1.** Pawl lock

_______________ **2.** Fly section

_______________ **3.** Butt end

_______________ **4.** Siderail

_______________ **5.** Bed section

_______________ **6.** Halyard

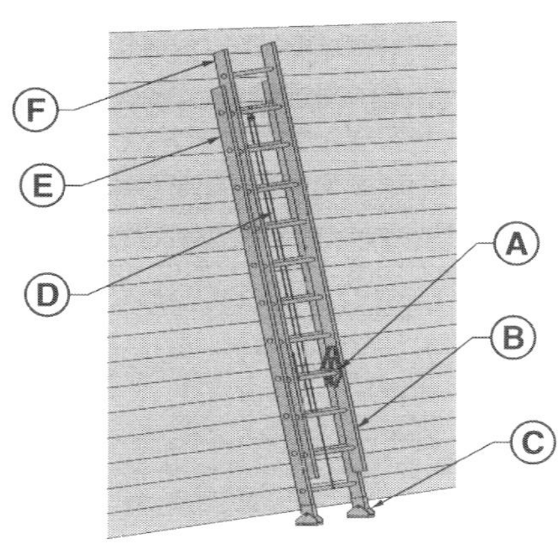

Pole Scaffold

_______________ **1.** Toeboard

_______________ **2.** Guardrail

_______________ **3.** Platform

_______________ **4.** Upright

_______________ **5.** Footing

_______________ **6.** Midrail

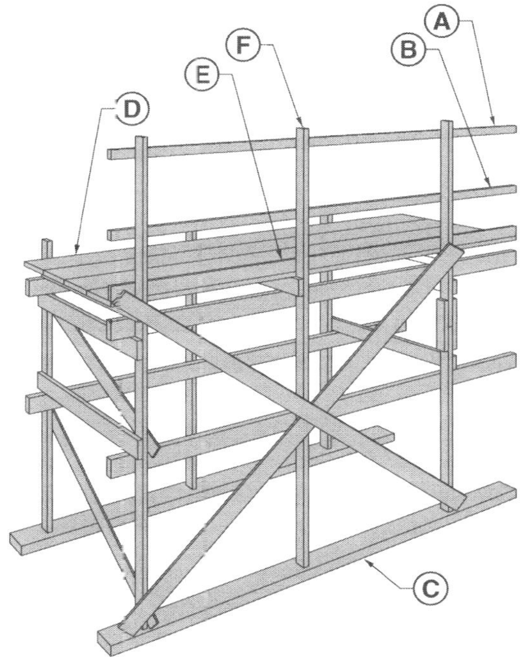

Sectional Metal-Framed Scaffold

_______________ **1.** Cleat

_______________ **2.** Planking

_______________ **3.** Bearer

_______________ **4.** Cross brace

_______________ **5.** Coupling tube

_______________ **6.** Hook-on ladder

_______________ **7.** Diagonal brace

_______________ **8.** Footing base plate

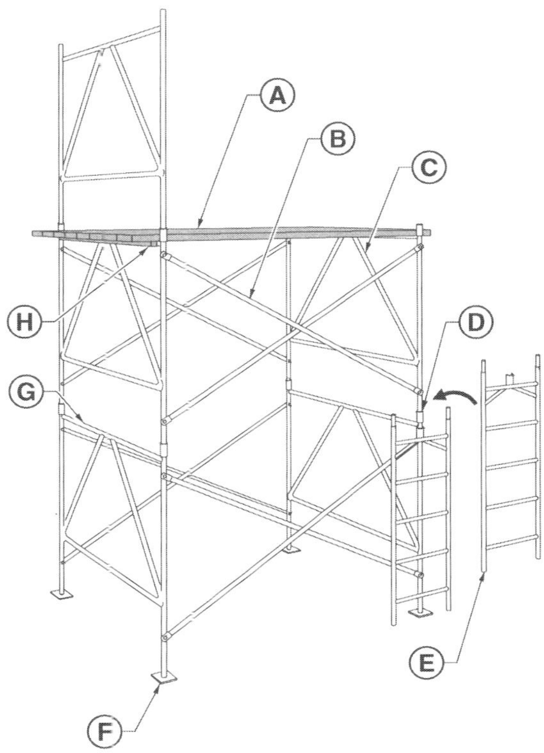

Two-Point Suspension Scaffold

_____ 1. Midrail

_____ 2. Lifeline

_____ 3. Platform

_____ 4. Toeboard

_____ 5. Toprail

_____ 6. Wire rope

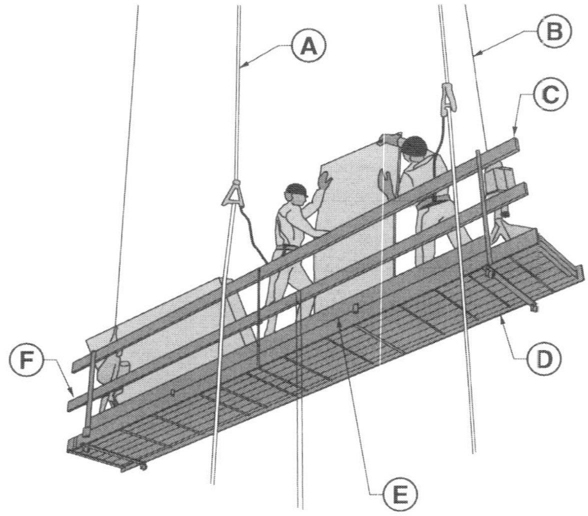

Activities

Extension Ladder Placement

Extension ladders are positioned against a structure on a 4:1 ratio. For every 4′ of working height, 1′ of space is required at the base. Working height (WH) is the distance from the ground to the top support. The top support is the area of a ladder that makes contact with a structure. For example, an extension ladder with its top support at a working height of 12′ requires the base to be placed 3′ from the structure (12′ ÷ 4 = 3′).

Determine the distance that the base must be placed away from the structure for the following working heights:

_____ 1. Working height = 20′

_____ 2. Working height = 32′

_____ 3. Working height = 10′

_____ 4. Working height = 18′

_____ 5. Working height = 26′

Projects

Contact tool and equipment manufacturers to request literature on their products. Review the product literature and compare the features of the tools and equipment. Develop a list of tools and equipment that an apprentice cement mason might purchase to perform basic concrete operations.

Concrete Principles

Chapter 3

Soil Conditions and Preparation

Name ______________________________ **Date** ____________________

True-False

T F **1.** The American Society for Testing and Materials and the American Association of State Highway and Transportation Officials classify soil in four major groups as determined by soil density.

T F **2.** Topsoil should always be removed from the area where concrete will be placed.

T F **3.** A micron is a unit of length equal to one-hundredth of a millimeter (0.01 mm).

T F **4.** A poorly-graded soil compacts to a higher density than a well-graded soil.

T F **5.** Soil density is expressed in pounds per cubic yard.

T F **6.** Soil is classified by the amount that can pass through a specific sieve size.

T F **7.** Moisture acts as a lubricant between soil particles.

T F **8.** In its natural state, topsoil does not contain any air or moisture in the voids between soil particles.

T F **9.** A fine-grained soil is a soil that consists mostly of sand and gravel.

T F **10.** A moisture-density curve is produced by plotting the dry density and moisture content values obtained from samples tested during a Proctor test.

T F **11.** A soil control curve is a chart that shows typical dry density and moisture content values for specific soil types.

T F **12.** The nuclear test is rarely used because it is time-consuming and produces questionable results.

T F **13.** A pneumatic backfill tamper is a vibratory rammer with a small round shoe.

T F **14.** A reversible vibratory plate produces 3000 lb to 15,600 lb of centrifugal force and can vibrate at a frequency of 3360 vpm to 7000 vpm.

T F **15.** The lugs of a sheepsfoot roller provide kneading action with vibratory force.

T F **16.** Walk-behind and ride-on vibratory smooth-drum rollers are ineffective on granular soil.

T F **17.** Overcompaction causes soil particles to spread out under pressure.

T F **18.** Granular soil, when used as disturbed backfill material, is effectively and economically compacted by vibration.

T F **19.** Compaction equipment should have a lift rating less than the desired lift thickness.

T F **20.** Carbon monoxide is heavier than air and can accumulate in low enclosed areas.

Multiple Choice

______ **1.** A ___ is a concrete, steel, or wood structural member embedded on end in the ground to support a load or to compact the soil.
A. column
B. pier
C. caisson
D. pile

______ **2.** ___ weathering is the decomposition of rock into soil particles by running water, freezing and thawing, and/or other natural means.
A. Chemical
B. Natural
C. Physical
D. Electrochemical

______ **3.** A ___ test is a soil test that measures and expresses attainable soil density and the effect of moisture on soil density.
A. slump
B. nuclear
C. Proctor
D. sand-cone

______ **4.** The Modified Proctor Test uses a ___ compared to the Standard Proctor Test.
A. greater dropping distance for a hammer
B. heavier hammer
C. greater number of soil layers
D. all of the above

______ **5.** ___ is any natural mineral material found on the surface of the earth except for embedded rock and organic plant and animal material.
A. Soil
B. Humus
C. Bedrock
D. none of the above

______ **6.** ___ is the process of condensing loose soil by applying energy.
A. Soil compaction
B. Soil displacement
C. Vibration
D. Consolidation

______ **7.** ___ is the uppermost layer of earth that supports plant growth.
A. Bedrock
B. Humus
C. Topsoil
D. Silt

______________ **8.** ___ is the ability of a material to support weight.
A. Soil swell
B. Compaction
C. Load-bearing capacity
D. Soil density

______________ **9.** ___ weathering is the decomposition of rock into soil particles by oxidation and/or natural acids.
A. Chemical
B. Natural
C. Physical
D. Electrochemical

______________ **10.** ___ soil consists mostly of sand and gravel with large visible particles.
A. Fine-grained
B. Cohesive
C. Gap-graded
D. Coarse-grained

______________ **11.** Soil ___ is the volume growth in soil after it is disturbed.
A. expansion
B. swell
C. augmentation
D. none of the above

______________ **12.** A ___ is a vertical steel H beam that is driven into the ground.
A. trench box
B. bench
C. caisson
D. soldier pile

______________ **13.** Soil compaction equipment is selected by considering ___.
A. soil type
B. job site conditions
C. soil compaction specifications
D. all of the above

______________ **14.** Vibratory plates are commonly specified for use on granular soil because of their ___.
A. dependability
B. economy
C. productivity
D. all of the above

______________ **15.** ___ soil is any soil consisting of cohesive and granular soil.
A. Gap-graded
B. Mixed
C. Cohesive
D. none of the above

______________ **16.** ___ compaction specifications allow the use of any equipment to achieve the specified Proctor density.
- A. Method
- B. End result
- C. Soil
- D. Project

______________ **17.** The methods used to protect against collapsing earth banks are specified in OSHA standards and are determined by the ___.
- A. depth of excavation
- B. type of structure under construction
- C. type and condition of the soil
- D. all of the above

______________ **18.** A ___ is a reinforced assembly consisting of two plates held apart by spacers used to shore the sides of a trench.
- A. waler
- B. trench box
- C. parapet
- D. soldier pile

______________ **19.** Excavated soil, machinery, and other materials must be at least ___′ away from the edge of an excavation.
- A. 2
- B. 4
- C. 8
- D. 10

______________ **20.** A(n) ___ is any construction cut, cavity, or depression in the surface of the earth formed by earth removal.
- A. trench
- B. excavation
- C. furrow
- D. pit

______________ **21.** A ramp, runway, ladder, or stair must be located within ___′ of an employee work area if a trench is 4′ or more in depth to provide a means of access or egress in case of an emergency.
- A. 10
- B. 15
- C. 20
- D. 25

______________ **22.** Soil particles with a large mass respond favorably to ___ vibration frequencies.
- A. high
- B. low
- C. neutral
- D. none of the above

_______________ **23.** Ride-on vibratory smooth-drum rollers produce 3000 lb to 9300 lb of centrifugal force and vibrate at a frequency of ___ vpm.
A. 1500
B. 2700
C. 3900
D. 4600

_______________ **24.** ___ is the maximum displacement from the neutral axis of a vibration wave to the outer edge of the wave.
A. Frequency
B. Vibration
C. Hertz
D. Amplitude

_______________ **25.** A ___ is a self-contained, self-advancing soil compaction machine that produces a high-frequency vibration that is transferred to the soil.
A. vibratory smooth-drum roller
B. pneumatic backfill tamper
C. vibratory sheepsfoot roller
D. vibratory plate

_______________ **26.** ___ is the use of wood or metal members to temporarily support soil, formwork, or construction materials.
A. Scaffolding
B. Shoring
C. Lagging
D. none of the above

_______________ **27.** The ___ test is a field soil test that measures the dry density of compacted soil.
A. nuclear
B. slump
C. sand-cone
D. none of the above

_______________ **28.** The theoretical point at which soil is at maximum density is the ___.
A. 100% Proctor density
B. Optimum moisture
C. Zero air void
D. Optimum density

_______________ **29.** OSHA regulations specify that banks more than ___′ high shall be shored or laid back to a stable slope, or some other equivalent means of protection shall be provided where employees may be exposed to moving ground or cave-ins.
A. 5
B. 10
C. 15
D. 20

_______________ **30.** The Proctor Test uses a soil sample of ___ cu ft.
A. 1⁄30
B. 1⁄3
C. 1⁄2
D. 1

_______________ **31.** Soil compaction can ___.
A. increase load-bearing capacity
B. reduce settling
C. reduce shrinkage
D. all of the above

_______________ **32.** ___ soil is created at one location and moved to another location.
A. Residual
B. Transported
C. Cohesive
D. Granular

_______________ **33.** Soil ___ is the number of soil particles present in a given volume of soil.
A. density
B. gradation
C. uniformity
D. composition

_______________ **34.** ___ is the grading of a soil sample based on the soil particle sizes present.
A. Density
B. Gradation
C. Uniformity
D. Composition

_______________ **35.** The maximum dry density attainable when a specific amount of compaction energy is applied at a certain moisture content is the ___.
A. 100% Proctor density
B. Optimum moisture
C. Zero air void
D. Optimum density

_______________ **36.** When conducting a sand-cone test, the ___ is used to control the hole diameter.
A. metal plate
B. scale
C. field density plate
D. all of the above

_______________ **37.** Vertical walls produced by benching cannot exceed ___′ in height.
A. 2
B. 2½
C. 3
D. 3½

_____ **38.** The internal spring system of a vibratory rammer produces approximately ___.
- A. 100 bpm to 250 bpm
- B. 250 bpm to 500 bpm
- C. 500 bpm to 800 bpm
- D. 800 bpm to 1000 bpm

_____ **39.** A common soil compaction method is ___ force compaction.
- A. impact
- B. static
- C. vibration
- D. all of the above

_____ **40.** Trench boxes are made from ___.
- A. steel
- B. concrete
- C. wood
- D. all of the above

_____ **41.** Method compaction specifications indicate the ___.
- A. soil compaction equipment
- B. lift
- C. number of passes required
- D. all of the above

_____ **42.** A ___ roller is a narrow, walk-behind rolling soil compaction machine with front and rear vibrating drums that have a pattern of protruding lugs.
- A. walk-behind vibratory
- B. vibratory sheepsfoot
- C. ride-on vibratory smooth-drum
- D. none of the above

_____ **43.** ___ cuts back trench walls into a step pattern.
- A. Sloping
- B. Excavating
- C. Benching
- D. Cutting

_____ **44.** ___ compaction is impact force compaction performed by dropping a heavy weight on a thick layer of uncompacted soil.
- A. Vibration force
- B. Static force
- C. Deep dynamic
- D. Roller

_____ **45.** Single-direction vibratory plates produce ___ of centrifugal force and vibrate at a frequency of 5600 vpm to 7000 vpm.
- A. 1000 lb to 1900 lb
- B. 2000 lb to 2900 lb
- C. 3000 lb to 4500 lb
- D. 4000 lb to 4900 lb

__________ **46.** A ___ roller is a self-advancing, articulated, rolling soil compaction machine with a smooth vibrating steel front drum and a smooth static steel rear drum.
A. walk-behind vibratory
B. vibratory sheepsfoot
C. ride-on vibratory smooth-drum
D. none of the above

__________ **47.** Making three passes with a rammer, four passes with a plate, or ___ passes with a roller typically achieves the required soil density if properly matched to the soil and soil moisture content is close to optimum.
A. three
B. four
C. five
D. six

Completion

__________ **1.** ___ is solid rock below the soil layer.

__________ **2.** The ___ is a soil classification system that indicates the quality of soil as a construction material.

__________ **3.** ___ is the soil moisture value at which maximum dry density is reached.

__________ **4.** A(n) ___ is a field soil test that measures moisture content and density of compacted soil with a portable nuclear gauge.

__________ **5.** A(n) ___ is a soil that remains in a compacted state after a weight is removed.

__________ **6.** A(n) ___ is an inorganic substance comprised of a solid crystalline chemical element or compound that is commonly extracted from the earth.

__________ **7.** Soil particles greater than 3″ are classified as ___.

__________ **8.** During operation, the lower end of a vibratory rammer moves up and down approximately ___″.

__________ **9.** ___ is fertile soil produced by decomposition of plant and/or animal matter.

__________ **10.** ___ is soil that remains near the site of the original decomposed rock.

__________ **11.** A(n) ___ is one trip over the soil with soil compaction equipment.

__________ **12.** When conducting a sand-cone test, a hole approximately 6″ in diameter and ___″ deep is dug in compacted soil.

__________ **13.** ___ is the continued application of soil compaction equipment after the desired density has been achieved.

__________ **14.** ___ force is the outward force produced by a rotating object.

__________ **15.** Soil particles with a(n) ___ mass respond favorably to high vibration frequencies.

______________________ **16.** The soil compaction machine component that produces high-frequency vibrations through unbalanced eccentric weights on a rotating shaft is the ___.

______________________ **17.** ___ soil consists of all cohesive or all granular soil.

______________________ **18.** The thickness of a layer of loose soil to be compacted is a(n) ___.

______________________ **19.** A(n) ___ is a narrow excavation that is 15′ wide or less made below the surface of the ground in which the depth is greater than the width.

______________________ **20.** ___ is the use of a portable protective device capable of withstanding forces from a cave-in.

______________________ **21.** ___ is the process of cutting back trench walls to an angle that eliminates the chance of collapse into the work area.

______________________ **22.** ___ remote control is a method of controlling machine functions using invisible radiation.

______________________ **23.** A horizontal support member used to hold trench sheet piling is a(n) ___.

______________________ **24.** ___ is planks used to retain earth on the side of a trench or excavation.

______________________ **25.** Vertical ___ uses opposing vertical structural members separated by screwjacks or hydraulic or pneumatic cylinders.

Short Answer

1. Excavations are commonly encountered on a job site. List five safety precautions and considerations that should be taken when working around or in excavations.

____

____

____

____

____

____

____

____

2. Soil compaction equipment includes vibratory rammers, pneumatic backfill tampers, vibratory plates, vibratory smooth-drum rollers, vibratory rollers, and vibratory sheepsfoot rollers. List these pieces of equipment along with their typical operating weights and the approximate number of blows or vibrations per minute for each.

____

____

____

____

____

____

Matching

Soil Particle Sizes

______________	**1.** Clay	**A.** Greater than 0.08″ up to and including 3″
______________	**2.** Silt	**B.** Up to and including 0.0002″
______________	**3.** Sand	**C.** Greater than 0.003″ up to and including 0.08″
______________	**4.** Gravel	**D.** Greater than 0.0002″ up to and including 0.003″

Soil Deposits

______________	**1.** Metamorphic rock	**A.** Formed from the solidification of molten lava
______________	**2.** Sedimentary rock	**B.** Changed in composition or texture by extreme heat, pressure, water, or chemicals
______________	**3.** Igneous rock	**C.** Formed from deposits such as sand, silt, and rock and shell fragments

Soil Particle Size Distribution

______	**1.** Uniform	**A.** Does not contain certain soil particle sizes
______	**2.** Gap-graded	**B.** Contains a broad range of soil particle sizes
______	**3.** Well-graded	**C.** Contains a limited range of soil particle sizes

Soil Compaction Methods

______	**1.** Static force	**A.** Uses weight from a heavy machine to squeeze soil particles together
______	**2.** Vibration force	**B.** Uses a machine that delivers a rapid succession of blows to the soil
______	**3.** Impact force	**C.** Uses a machine that delivers a high-frequency vibration to the soil

Vibratory Rammer

______ **1.** Gear

______ **2.** Shoe

______ **3.** Engine or motor

______ **4.** Guide handle

______ **5.** Centrifugal clutch

______ **6.** Connecting rod

______ **7.** Internal spring system

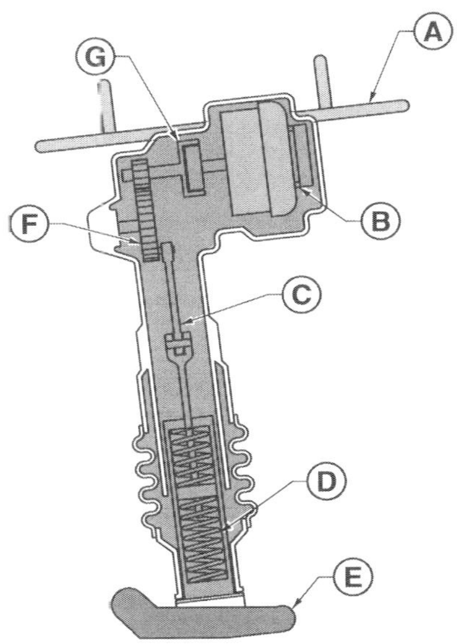

Proctor Test Equipment

_______________ **1.** Mold

_______________ **2.** Lifting knob

_______________ **3.** 5.5 lb hammer

_______________ **4.** Removable guide collar

_______________ **5.** Hammer guide sleeve

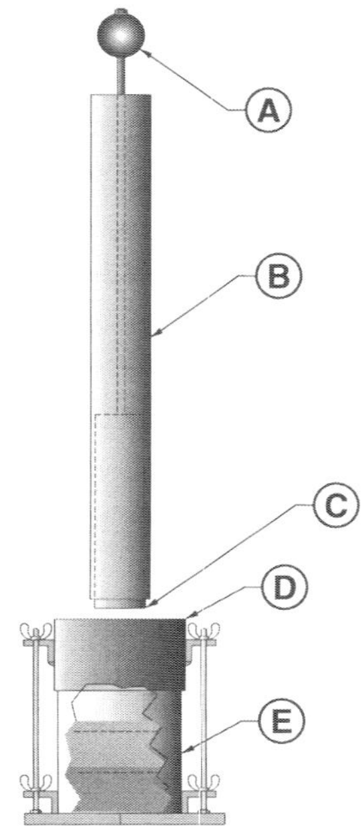

Vibratory Plate

_______________ **1.** Drive pulley

_______________ **2.** Gasoline or diesel engine

_______________ **3.** Exciter unit shaft

_______________ **4.** Base plate

_______________ **5.** Centrifugal clutch

_______________ **6.** Eccentric weight

_______________ **7.** Shock mount

_______________ **8.** Mount and bearing block

_______________ **9.** Exciter unit

_______________ **10.** Drive belt

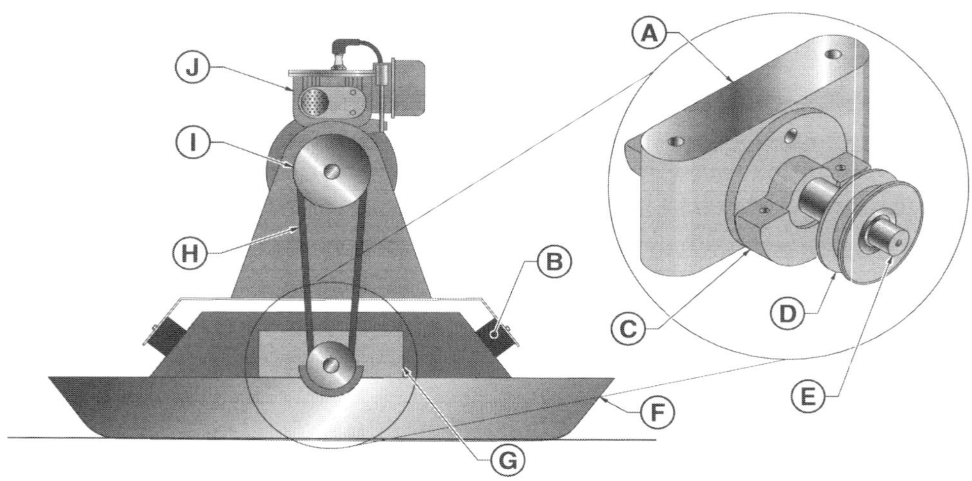

Activities

Trench Sloping

Per OSHA standards, trenches dug in different types of soil require different slopes to ensure worker safety. Trenches dug in highly cohesive soil must be sloped 53°; trenches in cohesive soil must be sloped 45°; trenches in granular soil must be sloped 34°. The approximate slope can be calculated by multiplying the depth of the trench by the cotangent of the slope angle. The cotangent of 53° is 0.7265, the cotangent of 45° is 1.0, and the cotangent of 34° is 1.4826. For example, the cutback distance of a 6′ deep trench dug in highly cohesive soil (53° slope) is 4.36′ (6′ × 0.7265 = 4.36′).

Determine the cutback distances of the following trenches:

____________________ **1.** 6′ deep trench in cohesive soil

____________________ **2.** 6′ deep trench in granular soil

____________________ **3.** 10′ deep trench in highly cohesive soil

____________________ **4.** 8′ deep trench in granular soil

____________________ **5.** 12′ deep trench in cohesive soil

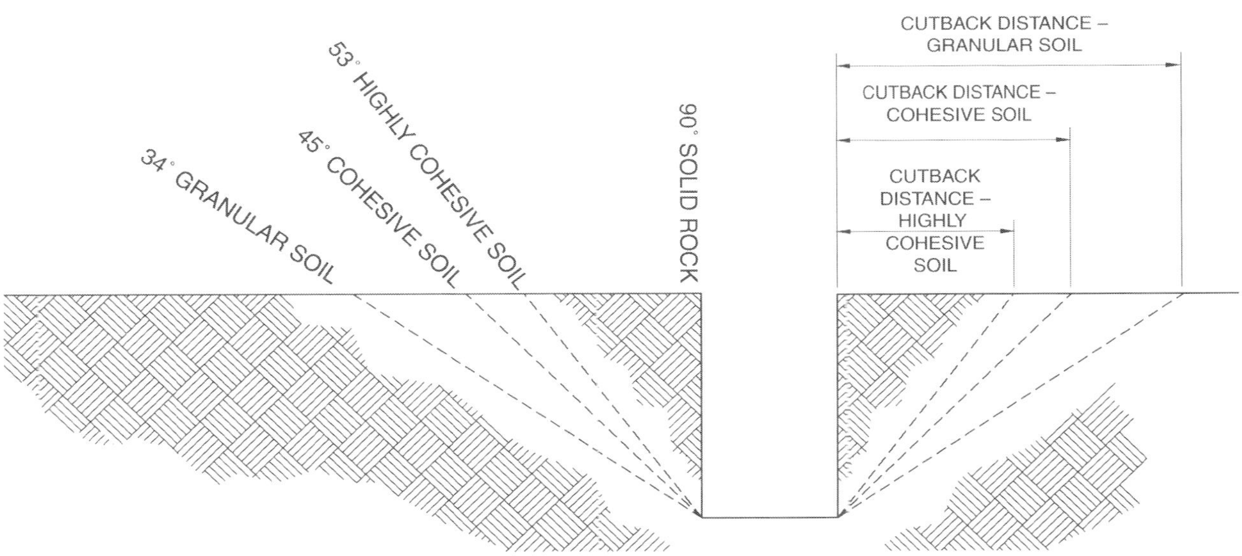

Projects

1. Different types of soil are predominant in different areas of a country. Various reasons may be cited for the predominance of the type of soil in a particular location, including glacial deposits and wind and water erosion. Identify the predominant type of soil in your area of the country and research how this type of soil may have become predominant.

2. Contact a local civil or geotechnical engineering firm regarding the types of soil tests they perform for a typical construction project. Identify challenges the firm may encounter in performing the tests and how they might address these challenges. Inquire about environmental tests they may perform including brownfield evaluation and stormwater management.

3. The methods used to protect workers against collapsing earth banks are specified in OSHA standards and are determined by the type and condition of the soil, depth of the excavation, level of the water table, type of structure under construction, and amount of space surrounding the excavation. Reference OSHA 29 CFR 1926.652, *Excavations* to learn more about the OSHA excavation standard.

Concrete Principles

Chapter 4

Flatwork

Name ______________________ Date ______________

True-False

T	F	**1.** Slipform paving eliminates the need for repeated erecting and stripping of forms, increases production, and produces a consistent, level, and durable surface.
T	F	**2.** In joist systems, the concrete for the slab is placed separately from the beams and girders.
T	F	**3.** Pans for forming joist systems are made from steel or fiberglass.
T	F	**4.** Flat slab and flat plate systems do not require the use of beams and girders.
T	F	**5.** A two-way joist is a floor slab system that has joists running at 45° angles to each other.
T	F	**6.** Approaches are built to twice the width of the driveway to allow for the tight turn radius of the rear wheels of a vehicle.
T	F	**7.** Walkways should slope ⅛″ to ¼″ per foot away from the building to allow for drainage.
T	F	**8.** Walkways that are set against a building should be at least 10″ to 12″ below the stoop or door sill.
T	F	**9.** The first step in site preparation is to excavate the topsoil to remove all organic matter from the site.
T	F	**10.** Front walks are usually 5′ wide.
T	F	**11.** Compaction increases the strength of the subgrade and provides the necessary support to prevent concrete from settling.
T	F	**12.** A base of compactible, drainable fill is not recommended over a vapor barrier.
T	F	**13.** Service walks are commonly 2′ to 5′ wide and are set about 2′ away from the foundation.
T	F	**14.** When preparing a site for concrete placement, fill material should be set in a single layer.
T	F	**15.** Walkways are commonly 4″ thick and require rebar in sections over which heavy loads must cross.
T	F	**16.** Welded wire reinforcement is represented by a long solid line in a section drawing.
T	F	**17.** Chairs are classified according to the rust protection they provide.

T F **18.** Five common rebar tying methods are single, wrap and snap, saddle (U-tie), wrap and saddle, or figure eight.

T F **19.** Welded wire reinforcement is only available in rolls.

T F **20.** A cutting torch should be used to cut epoxy-coated rebar.

T F **21.** Chairs are set perpendicular to the formwork.

T F **22.** Wire size for welded wire reinforcement is denoted by numbers and letters.

T F **23.** The load-carrying capacity of concrete is not affected by the elevation at which the reinforcement is set.

T F **24.** Wire for welded wire reinforcement may be smooth or deformed.

T F **25.** Deformed rebar has lugs that are rolled on during manufacturing and provide extra bonding between rebar and concrete.

T F **26.** Wired lap splices for rebar provide a stronger connection between rebar than a welded joint.

T F **27.** Lapped rebar splices are the least expensive and the easiest to make.

T F **28.** Dowels are designed to allow the space between sections of concrete to open and close during dimensional changes.

T F **29.** Dowels are not bonded to either side of adjacent sections of a concrete slab.

T F **30.** Nails should not be driven from the forms into stakes, as the nails are difficult to remove after concrete has been placed.

T F **31.** Improper setting of dowels restricts movement of a slab and may cause additional cracking.

T F **32.** Flat bulkheads are commonly used for thick slabs.

T F **33.** When forming a slab, stakes and braces should be cut flush with the forms to keep from interfering with screeding and finishing operations.

T F **34.** When applying a form-release agent, the second coat should be applied in the same direction as the initial coat.

T F **35.** Bulkheads are commonly used for large jobs where concrete placement may be halted for 30 min or more or where concrete placement must be placed on consecutive days to complete a job.

T F **36.** Concrete should be placed as close to its final position as possible to avoid segregation resulting from excessive movement.

T F **37.** Driveways used for commercial and industrial buildings that must support heavy loads are usually 4″ thick.

T F **38.** Keyed bulkheads create a tongue-and-groove joint.

T F **39.** Form-release agents should be applied to formwork after reinforcement is in position.

T F **40.** When placing concrete for a slab, start as close as possible to the concrete supply.

T F **41.** A single coat of form-release agent is adequate for most applications.

T F **42.** When placing concrete for a slab, the concrete should be placed in long continuous strips to permit easy access to sections as they are placed.

T F **43.** For small flatwork applications, consolidation can be accomplished using a tamper to push aggregate down into the mixture or by tapping the edge of the forms using a hammer.

T F **44.** Bull floating is typically performed in the same direction as the screeding operation.

T F **45.** Curing maintains the proper moisture content and temperature of concrete after placement to produce the desired characteristics of concrete.

T F **46.** If control joints are not used after initial curing, stress cracks appear and additional cracking occurs.

T F **47.** An expansion joint passes completely through the thickness of concrete.

T F **48.** Expansion joint material provides space for concrete to expand during temperature and moisture changes.

T F **49.** Air-entrained concrete should be floated before bleedwater evaporates.

T F **50.** Curing should begin as soon as possible after concrete has been finished.

T F **51.** Control joints relieve internal stresses and cause cracks to occur in a predetermined location.

T F **52.** Rebar should be placed across expansion joints to restrict the freedom of movement between building elements.

T F **53.** Floating or troweling mix water back into concrete results in the surface of the slab having a lower water-cement ratio than the bottom of the slab.

T F **54.** The higher the F-number, the flatter and more level the floor.

T F **55.** The F_L tolerance does not apply to inclined slabs or to slabs placed on unsupported formed surfaces.

T F **56.** The F-number system was developed in the 1960s as a method to control flat floor tolerances.

T F **57.** Hardeners are available only in powder form.

T F **58.** A wet screed is the preferred method for screeding flat floors as it is capable of achieving higher F-numbers than a vibratory truss screed.

T F **59.** Recent updates to ASTM E1155 state that parallel measurement lines for random traffic floor tolerances must be set more than 4′ from each other to prevent biases in test results.

T F **60.** Superflat floors are measured using a straightedge.

T F **61.** Power floating using pans on a flat floor can begin immediately after bull floating.

T F **62.** Superflat floors are measured using a profilograph.

Multiple Choice

______________ **1.** A(n) ___ is a surface paved with aggregate and concrete used to support vehicular traffic.
A. Jersey barrier
B. apron
C. concrete pavement
D. none of the above

______________ **2.** A ___ system is a concrete slab reinforced in two directions that uses drop panels or capitals to support the slab.
A. flat plate
B. flat slab
C. drop panel
D. slipform

______________ **3.** A ___ is a horizontal support member to which slab, floor, and ceiling materials are fastened.
A. beam
B. girder
C. waler
D. joist

______________ **4.** A ___ is a floor slab system that has cast-in-place joists running in one direction.
A. two-way joist
B. one-way joist
C. capital
D. flat plate system

______________ **5.** A(n) ___ floor slab is a concrete floor placed as a single, continuous member.
A. monolithic
B. two-course
C. bonded
D. unbonded

______________ **6.** A(n) ___ floor slab is composed of a base coat of concrete with high-strength mineral or metallic aggregate applied as a topping.
A. monolithic
B. two-course
C. bonded
D. unbonded

______________ **7.** Heavy-duty industrial floors require a(n) ___″ slab.
A. 4
B. 6
C. 8
D. 10

_______________ **8.** Garage floors subjected to heavy loads such as heavy machinery should be at least ___″ thick.
A. 4
B. 6
C. 8
D. 10

_______________ **9.** A garage slab is sloped ___ per foot to the entrance of the garage for drainage.
A. ⅛″ to ¼″
B. ¼″ to ½″
C. ½″ to ¾″
D. ¾″ to 1″

_______________ **10.** Driveway slabs for passenger cars are generally ___″ thick.
A. 4
B. 6
C. 8
D. 10

_______________ **11.** The standard width for a single residential driveway is ___′.
A. 10
B. 15
C. 20
D. 30

_______________ **12.** Patio slabs should be ___″ thick.
A. 4
B. 6
C. 8
D. 10

_______________ **13.** A ___ is a walkway that runs alongside the street and borders the building lot.
A. front walk
B. service walk
C. public sidewalk
D. none of the above

_______________ **14.** The standard width for a double commercial driveway is ___′.
A. 10
B. 15
C. 20
D. 30

_______________ **15.** Control joints should be placed no less than ___″ apart in 2′ wide walkways.
A. 20
B. 40
C. 60
D. 80

______________ **16.** A ___ is a short, large-diameter steel rod used to support the edges of two adjoining slabs.
- A. tie bar
- B. rebar
- C. dowel
- D. none of the above

______________ **17.** The temporary structure or mold used to retain and support concrete while it is setting and hardening is a ___.
- A. form
- B. brace
- C. trench
- D. none of the above

______________ **18.** Welded wire reinforcement is heavy-gauge wire joined in a grid and used to reinforce and increase the ___ strength of concrete.
- A. compressive
- B. shearing
- C. tensile
- D. yield

______________ **19.** Rebar is identified by markings located at one end of the bar representing the ___ of the rebar.
- A. type of steel
- B. grade
- C. diameter
- D. all of the above

______________ **20.** Welded wire reinforcement should be set no more than ___″ from the top surface of concrete.
- A. 1
- B. 2
- C. 3
- D. 4

______________ **21.** A yield strength of ___ psi should be assumed for rebar if a number or extra rib is not present.
- A. 2000
- B. 4000
- C. 20,000
- D. 40,000

______________ **22.** ___ strength is the load limit that a material will bend or stretch and still return to its original size and shape.
- A. Compressive
- B. Shearing
- C. Tensile
- D. Yield

_____________________ **23.** Rebar is spliced using the ___ method.
A. field-welded
B. mechanically-coupled
C. lapped
D. all of the above

_____________________ **24.** Fibers in fiber-reinforced concrete ___.
A. reduce cracking
B. increase impact capacity
C. provide tensile strength
D. all of the above

_____________________ **25.** Public sidewalks are commonly ___ wide.
A. 2′ to 3′
B. 3′ to 4′
C. 4′ to 5′
D. 5′ to 6′

_____________________ **26.** Tie bars are commonly made from No. 4 or No. 5 rebar and are ___″ in length.
A. 20
B. 30
C. 40
D. 48

_____________________ **27.** A ___″ diameter by 18″ dowel should be placed 12″ OC for 9″ to 11″ thick slabs.
A. ½
B. ¾
C. 1
D. 1¼

_____________________ **28.** Vapor barrier joints are lapped at least ___″.
A. 2
B. 4
C. 6
D. 8

_____________________ **29.** Patios should be pitched ___ per foot away from the building.
A. ⅛″ to ¼″
B. ¼″ to ½″
C. ½″ to ¾″
D. ¾″ to 1″

_____________________ **30.** The first two numbers of a welded wire reinforcement specification denote wire ___.
A. size
B. spacing
C. finish
D. none of the above

______________ **31.** ___ is a tying method used for tying reinforcement in concrete mats that are to be lifted and placed in position with a crane.

A. Wrap and snap
B. Saddle
C. Wrap and saddle
D. Figure eight

______________ **32.** A ___ is a walkway that extends from a driveway or sidewalk to the rear entrance of a building.

A. front walk
B. service walk
C. public sidewalk
D. none of the above

______________ **33.** When forming a slab, intermediate stakes should be spaced approximately ___ apart.

A. 1′ to 3′
B. 3′ to 5′
C. 6′ to 8′
D. 8′ to 10′

______________ **34.** A form-release agent is a substance that ___.

A. allows forms to release cleanly from hardened concrete
B. protects the forms
C. aids in producing a hard and stain-free concrete surface
D. all of the above

______________ **35.** A ___ is a wood member installed inside or at the end of a concrete form to prevent concrete flow from coming into a section or out of the ends of the forms.

A. keystock
B. screed
C. corner stop
D. bulkhead

______________ **36.** A(n) ___ joint is a joint or discontinuity in concrete resulting from a delay in placement.

A. isolation
B. construction
C. cold
D. expansion

______________ **37.** A(n) ___-release agent is a form-release agent commonly made from paraffin wax that prevents concrete from sticking to the forms.

A. chemical
B. barrier
C. combination
D. inert

_______________ **38.** Hand floating or darby floating prepares a surface for troweling by ___.
A. removing small imperfections, humps, and depressions
B. embedding coarse aggregate below the surface
C. consolidating mortar at the surface
D. all of the above

_______________ **39.** When using a power trowel to finish a slab, each successive finishing pass should overlap the previous pass by ___ the trowel width to ensure complete slab coverage and to minimize surface imperfections.
A. one-fourth
B. one-third
C. one-half
D. three-fourths

_______________ **40.** Check rodding is performed perpendicular to the direction of screeding with additional check rodding passes made at a ___° angle to previous passes.
A. 30
B. 45
C. 60
D. 90

_______________ **41.** When determining floor flatness, test sections less than ___′ wide must use the diagonal layout method.
A. 10
B. 15
C. 20
D. 25

_______________ **42.** Curing should continue for a minimum of ___ days to allow adequate strength gain before concrete is exposed to weather.
A. two
B. five
C. seven
D. nine

_______________ **43.** Wire ties are commonly made from No. ___ black annealed wire and are available in precut lengths.
A. 4
B. 8
C. 16
D. 24

_______________ **44.** A ___ level is a surveying instrument that rotates vertically and horizontally and consists of a telescope and a leveling vial.
A. carpenter's
B. laser transit
C. transit
D. torpedo

__________ **45.** A general rule for concrete slabs is to place control joints about ___ times (in feet) the slab depth (in inches).

A. 2½
B. 4
C. 5½
D. 7

__________ **46.** A joint is a designed crack in concrete used primarily to ___.

A. allow free movement of a slab
B. reduce stress
C. minimize cracking
D. all of the above

__________ **47.** Control joints will range from 10% to 33% of the slab thickness, depending on ___.

A. mix design
B. cutting method
C. timing of the cut
D. all of the above

__________ **48.** Edging ___.

A. consolidates and shapes the concrete edge
B. compacts concrete next to the form
C. produces a radiused corner that resists breaking and chipping
D. all of the above

__________ **49.** Tie bars are usually spaced in concrete at ___″ intervals.

A. 24
B. 30
C. 36
D. 42

__________ **50.** Settlement in slabs is caused by ___.

A. variations in soil type
B. uneven loading on the foundation
C. uneven soil compaction
D. all of the above

__________ **51.** Bleedwater can be reduced by ___.

A. adding fly ash
B. increasing the cement content 25 lb to 50 lb per cubic yard of concrete
C. introducing or increasing air-entraining admixtures
D. all of the above

__________ **52.** A control joint should be cut within ___ hr of finish troweling of the section.

A. 2
B. 5
C. 8
D. 12

________________ **53.** A(n) ___ joint is a longitudinal joint used on driveways and highway decks to eliminate random cracking of concrete caused by warpage.

A. expansion
B. control
C. construction
D. warping

________________ **54.** A(n) ___ joint is used where two successive placements of concrete meet, across which a bond is maintained between the placements.

A. expansion
B. control
C. construction
D. isolation

________________ **55.** When testing a floor for flatness, measurements must be taken within ___ hr of placement, and the contractor must be notified of the results within 72 hr of testing.

A. 12
B. 24
C. 36
D. 48

________________ **56.** The flatness number (___) measures the flatness of a concrete floor.

A. F_F
B. F_L
C. F_{min}
D. F_{max}

________________ **57.** When determining floor flatness, the minimum dimensions of test sections are ___′.

A. 4
B. 8
C. 12
D. 16

________________ **58.** When using a vacuum dewatering system, approximately ___ of the initial mix water is removed from the slab in approximately 30 min.

A. 5% to 10%
B. 10% to 12%
C. 15% to 25%
D. 30% to 40%

________________ **59.** ___ is the least acceptable measurement for commercial floors.

A. $F_F 6/F_L 40$
B. $F_F 13/F_L 10$
C. $F_F 22/F_L 15$
D. $F_F 40/F_L 6$

________________ **60.** A Dipstick® is a floor profiling device that measures the floor at ___ points, each spaced 1′ apart.

A. two
B. three
C. four
D. six

__________ **61.** Dewatering ___.
A. increases concrete strength
B. reduces curing shrinkage
C. increases wear-resistance
D. all of the above

__________ **62.** A bull float is only capable of achieving a flatness number of ___.
A. F_F 5 to F_F 9
B. F_F 7 to F_F 10
C. F_F 17 to F_F 20
D. F_F 26 to F_F 30

__________ **63.** ___ floors are considered very flat.
A. F_F 12/F_L 30
B. F_F 13/F_L 10
C. F_F 40/F_L 25
D. F_F 50/F_L 30

__________ **64.** Superflat floors are typically measured with the ___ system.
A. F_F
B. F_L
C. F_{min}
D. F_{max}

__________ **65.** Superflat floors require a measurement of at least ___.
A. F_{min} 50
B. F_{max} 100
C. F_{min} 100
D. F_F 50

__________ **66.** Spacing of construction joints on superflat floors is limited to no more than ___′.
A. 6
B. 12
C. 16
D. 20

__________ **67.** A(n) ___ joint is a groove made in a horizontal concrete surface to create a weakened plane and control the location of cracking.
A. expansion
B. isolation
C. control
D. construction

__________ **68.** The F-number system was developed by the ___.
A. Canadian Standards Association
B. American Society for Testing and Materials
C. American Concrete Institute
D. all of the above

______ **69.** When trying to achieve high F-numbers on flat floors, the maximum span allowed for edge forms is ___′.

A. 10
B. 30
C. 50
D. 70

______ **70.** A(n) ___ joint separates adjoining sections of concrete to allow for movement caused by expansion and contraction of the slabs.

A. expansion
B. control
C. construction
D. warping

______ **71.** The levelness number (___) measures the tilt of a concrete floor or the change in elevation across the surface.

A. F_F
B. F_L
C. F_{min}
D. F_{max}

______ **72.** To correct unevenness in a concrete slab, restraightening should be performed at a ___° angle to placement, from both sides of the slab.

A. 30
B. 45
C. 60
D. 90

Completion

______ **1.** A(n) ___ is a thickened area over a column.

______ **2.** A(n) ___ system is a flat slab system used for light loads, such as office buildings or apartments, that does not use drop panels or capitals.

______ **3.** In general, industrial and commercial floors are ___″ thick.

______ **4.** A(n) ___ is a support structure made from metal, plastic, or precast concrete used to provide an accurate, consistent spacing between welded wire reinforcement or rebar and subgrade.

______ **5.** A square prefabricated pan form that is nailed in position to form a two-way joist system is a(n) ___ pan.

______ **6.** Most garage and basement floor slabs are ___″ thick.

______ **7.** A(n) ___ floor slab is a floor slab in which a portion of the original slab is removed and fresh concrete is reapplied over the remaining slab.

______ **8.** Driveway slabs that abut garage floors should have a finished surface that is ___″ lower than the garage floor to prevent water from entering the garage.

______________ **9.** Driveways can slope toward the garage if necessary, but a(n) ___ must be installed at the joint between the garage and the driveway.

______________ **10.** Control joints should be placed no less than ___″ apart in walkways 3′ or wider.

______________ **11.** ___ matter is material such as grass and roots of trees and shrubs.

______________ **12.** The ___ is the section(s) of concrete extending from the sidewalk to the building.

______________ **13.** A(n) ___ is a waterproof material that prevents ground moisture from penetrating into the slab.

______________ **14.** The section of concrete between the street and the sidewalk is the ___.

______________ **15.** A(n) ___ is a walkway that extends from a driveway or public sidewalk to the front entrance of a building.

______________ **16.** ___ is a steel bar that interlocks with concrete to increase the tensile strength of concrete.

______________ **17.** Control joints should be provided at a maximum of ___′ intervals in both directions in a patio slab.

______________ **18.** Two common protective coatings for rebar are epoxy and ___.

______________ **19.** A(n) ___ is an exterior concrete slab constructed adjacent to a building and used as an extension of the living or work area.

______________ **20.** A(n) ___ saw is used to cut epoxy-coated rebar.

______________ **21.** ___ concrete is a concrete mixture that uses glass, metal, or plastic fibers mixed with concrete to provide extra strength.

______________ **22.** Aggregate ___ is a method of relieving stresses on concrete by preventing uneven up-and-down movement between two slabs.

______________ **23.** A(n) ___ is a short piece of rebar used to join adjacent slabs or concrete sections.

______________ **24.** Regardless of slab depth or dowel diameter, dowels should be spaced ___″ apart, center to center.

______________ **25.** ___ is the entire system of support for fresh concrete, including forms, hardware, and bracing.

______________ **26.** The leveling process of formed concrete surfaces using guides and a straightedge or a vibratory truss screed is ___.

______________ **27.** A(n) ___ level is a metal or wood frame containing one or more clear vials containing fluid and an air bubble that indicates levelness by the location of the bubble in the vial(s).

______________ **28.** A(n) ___-release agent is a form-release agent made from fatty acids and petroleum.

________________ **29.** A(n) ___ is a stable reference point marked with the elevation above mean sea level from which differences in elevation around the job site are measured.

________________ **30.** A type of form-release agent that uses both a barrier-release agent and a chemical-release agent is a(n) ___-release agent.

________________ **31.** A(n) ___ floor is a floor that is extremely flat.

________________ **32.** Concrete should be cured for ___ days before being exposed to traffic or load conditions.

________________ **33.** The groove depth of a tooled control joint should be approximately ___ the slab thickness.

________________ **34.** During initial curing, concrete shrinks approximately ___″ per 100′.

________________ **35.** ___ is the chemical reaction of water and cement that bonds molecules, resulting in hardening of the concrete mixture.

________________ **36.** A(n) ___ construction joint is a construction joint that also acts as a control joint.

________________ **37.** ___ is water that is displaced to the surface of fresh concrete by aggregate sinking to the bottom of the concrete mixture.

________________ **38.** A(n) ___ is a groove formed into fresh concrete that is used to interlock concrete structures placed at different times.

________________ **39.** A(n) ___ traffic floor is a floor on which traffic travels the same path repeatedly.

________________ **40.** When determining floor flatness, the minimum area of a test section is ___ sq ft.

________________ **41.** Floor slope or levelness is measured across the floor over a distance of ___′.

________________ **42.** A(n) ___ is preferred for achieving flat floors as it has a flat blade surface and can achieve a high F-number of F_F25 to F_F35.

Short Answer

1. Flatwork procedures include formwork, placement, finishing, and curing of concrete. List six objectives that must be met to achieve high-quality flatwork.

____

____

____

____

____

2. Summarize the procedure for forming a slab.

3. Summarize the procedures for sawing and tooling a control joint.

4. List four general principles that should be followed when placing concrete flatwork.

Matching

Chair Classes

______	**1.** Class 1	**A.** Made from plain carbon steel and provide no rust protection
______	**2.** Class 2	**B.** Made entirely from plastic or have plastic-covered feet or legs
______	**3.** Class 3	**C.** Made entirely from stainless steel or have stainless steel feet

Floor Classifications

______	**1.** Class 1	**A.** Institutional and commercial applications
______	**2.** Class 2	**B.** Bonded two-course floors subject to heavy traffic
______	**3.** Class 3	**C.** Residential surfaces mainly with floor coverings
______	**4.** Class 4	**D.** Light-duty industrial floors for manufacturing, processing, and warehousing
______	**5.** Class 5	**E.** Narrow-aisle, high-bay warehouses; television studios
______	**6.** Class 6	**F.** Offices and churches, usually with floor covering; decorative.
______	**7.** Class 7	**G.** Industrial floors subject to heavy traffic; may be subject to impact loads
______	**8.** Class 8	**H.** Unbonded toppings – freezer floors on insulation, old floors, or where construction schedule dictates
______	**9.** Class 9	**I.** Exterior walkways, driveways, garage floors, and sidewalks

Curbs

______	**1.** Barrier	**A.** Reduced height to allow traffic to pass over
______	**2.** Jersey barrier	**B.** Redirects a vehicle traveling at impact speeds below 45 mph
______	**3.** Mountable	**C.** Redirects a vehicle traveling at high speeds

Chairs

__________ 1. Continuous high chair

__________ 2. Individual high chair

__________ 3. Joist chair upper

__________ 4. Slab bolster

__________ 5. Beam bolster upper

__________ 6. Joist chair

__________ 7. Continuous high chair upper for metal deck

__________ 8. Slab bolster upper

__________ 9. Continuous high chair upper

__________ 10. Beam bolster

__________ 11. High chair for metal deck

__________ 12. Individual bar chair

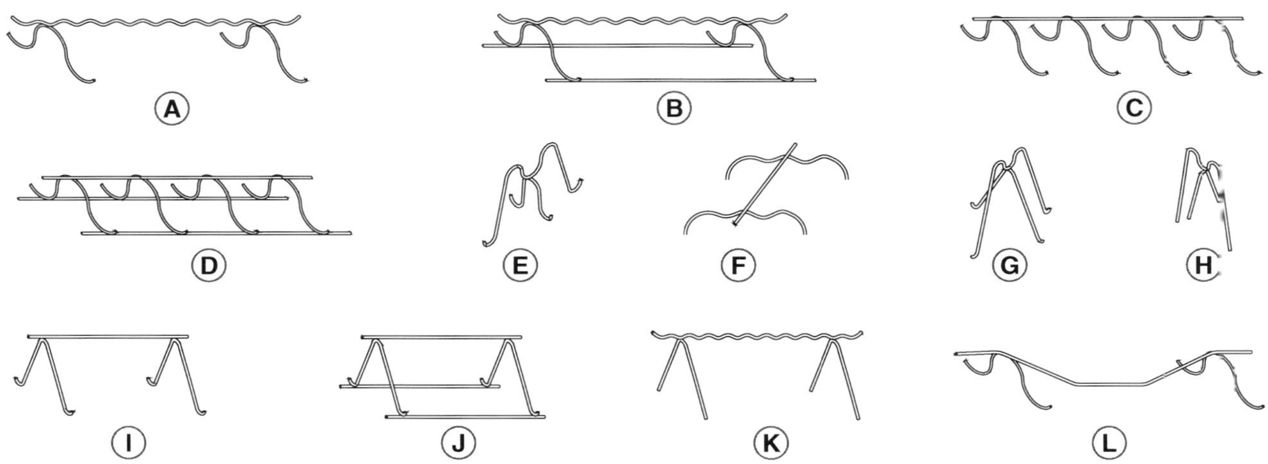

Rebar Tying Methods

__________ 1. Wrap and saddle

__________ 2. Single

__________ 3. Figure eight

__________ 4. Wrap and snap

__________ 5. Saddle (U-tie)

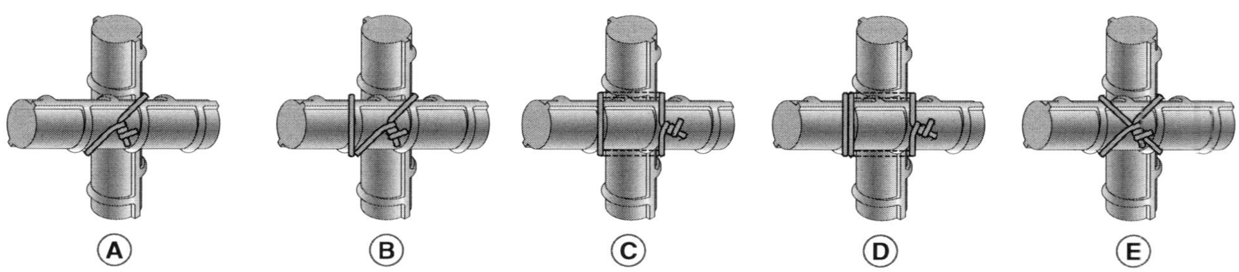

Activities

Driveway Slope

Driveways are typically sloped ⅛″ to ¼″ per foot to provide adequate drainage of rainwater. The amount of slope must be taken into consideration when setting up driveway formwork. Calculating the difference in elevation between the edge forms is based on the following formula:

$$E_D = \frac{S \times R}{12}$$

where

E_D = elevation difference (in in.)
S = slope (in in. per foot)
R = width (in in.)
12 = constant

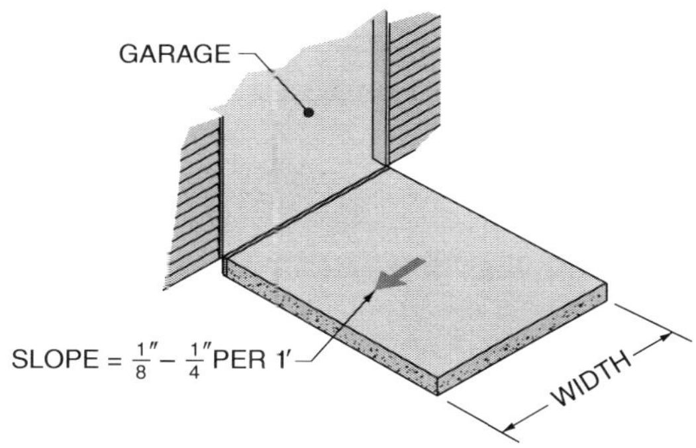

For example, determine the elevation difference between edge forms for a 10′ (120″) wide residential driveway sloped to one side at ¼″ (0.25″) per foot.

$$E_D = \frac{S \times R}{12}$$

$$E_D = \frac{0.25 \times 120}{12}$$

$$E_D = \frac{30}{12}$$

E_D = 2.5″ = **2½″**

Determine the difference in elevation between edge forms for the following driveways:

________________ **1.** 20′ wide residential driveway sloped to one side at ¼″ per foot

________________ **2.** 15′ wide commercial driveway sloped to one side at ¼″ per foot

________________ **3.** 30′ wide commercial driveway sloped to one side at ⅛″ per foot

________________ **4.** 10′ wide residential driveway sloped to one side at ¼″ per foot

Projects

1. Obtain a copy of a floor plan for a residential structure with an attached garage. Identify exterior doors on the structure, including doors leading from the garage to the outside. Using the appropriate scale, lay out all walkways for the structure, including the front walk and service walks, and note slopes that will be necessary to provide proper drainage. Lay out a driveway for the structure and determine the best direction for sloping the driveway.
2. Obtain a copy of the print specifications for a commercial structure and identify areas related to concrete work including formwork, reinforcement requirements, form-release agents, concrete mixtures, and concrete quality.

Concrete Principles

Chapter 5

Concrete Structures

Name ______________________ **Date** ______________

True-False

T	F	**1.** A pile driver is heavy construction equipment that uses a drop hammer, mechanical hammer, or vibratory hammer to drive piles into the ground.
T	F	**2.** A grouped pile is multiple friction piles that are driven in close arrangement.
T	F	**3.** A non-load-bearing wall supports only its own weight.
T	F	**4.** Friction piles must be driven down to load-bearing soil.
T	F	**5.** A large panel form is a wall form constructed of many small panels bolted together.
T	F	**6.** Control joints are formed vertically in walls by attaching a beveled strip of wood, metal, rubber, or plastic to the sheathing of the wall form.
T	F	**7.** When placing control joints for the upper floors of structures, the joints should be placed flush with the jamb line of door or window openings.
T	F	**8.** A cleanout door is placed at the base of a column form to clean debris from the form before concrete placement.
T	F	**9.** A ganged panel form is constructed in large prefabricated units.
T	F	**10.** Shearing stresses affect the horizontal axis of a beam.
T	F	**11.** When placing control joints for the first floor of a structure, the joints should be placed flush with the jamb line of door or window openings.
T	F	**12.** Horizontal stresses have a greater effect on wood beams than on concrete beams.
T	F	**13.** Static loads are variable and cause excessive stress on a foundation.
T	F	**14.** Horizontal shear affects the horizontal (transversal) axis of a beam.
T	F	**15.** Foundation walls are commonly 10″ or 12″ wide but can be wider depending on the application.
T	F	**16.** A footing should never be less than 6″ thick.
T	F	**17.** The bottom of a pier footing must extend below the frost line and rest on firm soil.
T	F	**18.** The footing depth should equal twice the foundation wall thickness.
T	F	**19.** Vapor barriers are often placed over the ground under crawl space foundations to prevent moisture buildup in the crawl space.
T	F	**20.** The bottom of a grade beam does not need to extend below the frost line.

T F **21.** A slab-on-grade foundation is placed directly on the ground.

T F **22.** A spread footing is not required with a rectangular foundation.

T F **23.** A short span of a bridge deck consists of a highway deck that rests directly on the piers.

T F **24.** Riser height is determined by dividing the number of risers by the total rise of the stairway.

T F **25.** Slipform-paving equipment is concrete forming equipment that is moved horizontally or vertically as the concrete is placed.

T F **26.** When attaching a form liner, the outer edges of the liner should be attached first and worked toward the center to prevent buckling.

T F **27.** An expansion joint is placed between the stairway and foundation wall to prevent cracking caused by movement of the stairway.

T F **28.** To achieve high-quality concrete in precast members, a high water-cement ratio should be used.

T F **29.** Floor slab and beam connections for precast members are made using angles and welding plates.

T F **30.** Precast beams that rest on the tops of columns are secured with steel angles welded to post-tensioning bars that extend from the ends of the beams.

Multiple Choice

______________________ **1.** A(n) ___ is a concrete, steel, or wood structural member embedded on end in the ground to support a load or to compact the soil.
A. abutment
B. caisson
C. pile
D. column

______________________ **2.** A ___ pile relies on surface friction with the soil to support an imposed load.
A. grouped
B. friction
C. bearing
D. all of the above

______________________ **3.** An offset where the vertical end of a wall panel fits into the recessed channel of a column is formed with a(n) ___.
A. chamfer strip
B. curtain wall
C. offset strip
D. recess strip

_______________ **4.** ___ is a method of prestressing reinforced concrete in which the tendons in the structural member are tensioned before the concrete has hardened.
A. Post-tensioning
B. Pretensioning
C. Precasting
D. none of the above

_______________ **5.** A ___ is a PVC, rubber, or stainless-steel barrier used to prevent the passage of liquid or gas under pressure through a joint in a concrete slab or wall.
A. waterstop
B. chamfer strip
C. recess strip
D. bulkhead

_______________ **6.** The point at which a portion of the pile head is removed after the pile has been driven into the proper position is the pile ___.
A. butt
B. foot
C. cutoff
D. shoe

_______________ **7.** A ___ is a metal cap placed on top of a pile head to receive the blow from a pile driver and protect the pile from damage.
A. pile head
B. pile butt
C. driving shoe
D. driving head

_______________ **8.** A(n) ___ insert consists of a T-bar anchor and a recess former supported by a base.
A. encasement ball
B. clutch-type
C. tilt-up
D. lift plate

_______________ **9.** A(n) ___ strip is a narrow strip of wood ripped at a 45° angle that produces a beveled edge on a concrete surface.
A. bevel
B. chamfer
C. recess
D. angle

_______________ **10.** The vertical structural member used to support compressive loads is a ___.
A. beam
B. girder
C. column
D. none of the above

_______________ **11.** A common column shape is ___.
A. square
B. round
C. rectangular
D. all of the above

______________ **12.** A ___ is a horizontal member, smaller than a girder, that supports a bending load over a span.

A. column
B. beam
C. spandrel
D. footing

______________ **13.** Commercial slabs are either ___ thick depending on the intended traffic load.

A. 2″ or 4″
B. 4″ or 6″
C. 6″ or 8″
D. 8″ or 10″

______________ **14.** The vertical load of a building is carried by a ___ wall.

A. curtain
B. load-bearing
C. non-load-bearing
D. none of the above

______________ **15.** Control joints for walls should be within ___ of a corner.

A. 2′ to 4′
B. 8′ to 10′
C. 10′ to 15′
D. 20′ to 25′

______________ **16.** Minimum crawl space foundation depths range from ___.

A. 6″ to 12″
B. 12″ to 18″
C. 18″ to 24″
D. 24″ to 30″

______________ **17.** Grade beams must be at least ___″ thick and 14″ deep.

A. 4
B. 6
C. 8
D. 10

______________ **18.** A ___ footing is a rectangular prism of concrete larger in lateral dimensions than the column or wall it supports that distributes the load of a column or wall to the subgrade.

A. square pier
B. combined
C. tapered pier
D. spread

______________ **19.** A ___ foundation is used in residential structures consisting of 9′ high walls to allow space for mechanical systems without sacrificing finished ceiling height.

A. crawl space
B. stepped
C. battered
D. full basement

______________ **20.** A ___ is a reinforced concrete beam placed on ground level and supported by piles or piers.

A. ground beam
B. grade beam
C. transversal beam
D. all of the above

______________ **21.** The vertical footing of a stepped foundation must be at least 6″ thick and no higher than ___ the distance between two horizontal steps.

A. one-fourth
B. one-third
C. one-half
D. three-fourths

______________ **22.** A(n) ___ foundation is a continuous slab of concrete, usually reinforced, laid over soft ground or where heavy loads must be supported to form a foundation.

A. mat
B. slab-on-grade
C. raft
D. battered

______________ **23.** A ___ foundation is a monolithic structural support consisting of a wall with a vertical exterior face and a sloping interior face.

A. mat
B. slab-on-grade
C. raft
D. battered

______________ **24.** A common highway pier footing shape is ___.

A. tapered
B. circular
C. stepped
D. all of the above

______________ **25.** A ___ foundation is a continuous footing with a slab-like shape that can be placed monolithically or as a separate footing and foundation and can support an array of columns in several rows in each direction.

A. mat
B. slab-on-grade
C. raft
D. battered

______________ **26.** A ___ is a 1″ or 2″ thick level piece of dimensional lumber formed to hold the building lines and to show the exact boundaries of a building.

A. corner stake
B. batterboard
C. building setback
D. none of the above

_______________ **27.** Grade beams should extend a minimum of ___″ above the finish grade in average soil conditions.
A. 6
B. 8
C. 10
D. 12

_______________ **28.** A(n) ___ is a cut or groove made by a saw blade in the upper edge of batterboards.
A. kerf
B. uplift
C. column
D. none of the above

_______________ **29.** ___ allow vehicular traffic to merge smoothly into highway traffic or onto a feeder roadway.
A. Overpasses
B. Approaches
C. Medians
D. none of the above

_______________ **30.** A ___ is a short section of wall attached at either end of an abutment and at a slight angle to the abutment.
A. pier cap
B. pier
C. wing wall
D. substructure

_______________ **31.** A large load-bearing surface and a direct support for a highway superstructure is a(n) ___.
A. wing wall
B. pier cap
C. abutment
D. substructure

_______________ **32.** ___ is the vertical distance from one floor level to the next level in a stairway.
A. Total rise
B. Total run
C. Tread depth
D. Stair height

_______________ **33.** The surface of a casting bed must be ___.
A. free from defects
B. level
C. smooth
D. all of the above

_______________ **34.** The ___ wall is the back wall of an abutment.
A. wing
B. stem
C. head
D. none of the above

______________ **35.** Manufactured form liners are made from ___.
A. rubber
B. polystyrene
C. PVC
D. all of the above

______________ **36.** Form liners are attached to forms with ___.
A. staples
B. nails
C. waterproof adhesives
D. all of the above

______________ **37.** A ___ plate is a steel plate connector that contacts connectors in other precast members after each member is placed.
A. matching
B. connecting
C. precast
D. job

______________ **38.** ___ must be used as a casting bed if panels are cast outside the building.
A. A temporary concrete slab
B. Well-compacted fill
C. A wood platform
D. all of the above

______________ **39.** ___ are used to raise precast members into position.
A. Anchors
B. Inserts
C. Lifting units
D. all of the above

______________ **40.** Tilt-up inserts include ___ that are embedded in precast tilt-up members.
A. anchors
B. lift plates
C. inserts
D. all of the above

______________ **41.** The ___ is the vertical wall of an abutment that is supported by a footing or other load-bearing surface.
A. wing wall
B. stem wall
C. pier cap
D. substructure

______________ **42.** A(n) ___ load is the weight of a single stationary body or the combined weights of all stationary bodies in a structure.
A. live
B. static
C. imposed
D. unimposed

_______________ **43.** The projection of the footing should be at least ___ the foundation wall thickness to each side of the foundation wall.
A. one-fourth
B. one-third
C. one-half
D. three-fourths

_______________ **44.** A ___° taper angle from the horizontal should be maintained on a tapered pier footing to provide a safety margin based on a 45° shear stress angle.
A. 15
B. 30
C. 45
D. 60

_______________ **45.** A ___ is the primary support for a structure through which imposed loads are transmitted to the ground.
A. footing
B. grade beam
C. foundation
D. column

_______________ **46.** A ___ is the portion of a foundation that spreads and transmits loads directly to piles or to the soil.
A. footing
B. column
C. keyway
D. pier

_______________ **47.** The ___ is the main support for a highway system.
A. superstructure
B. stem wall
C. wing wall
D. substructure

Completion

_______________ **1.** A(n) ___ is a pile that is driven down to a load-bearing soil such as bedrock.

_______________ **2.** A(n) ___ caisson is a flared caisson that provides a greater load-bearing surface at the base than a pile.

_______________ **3.** The pile ___ is the upper surface of a precast pile in its final position.

_______________ **4.** The metal cylindrical shell that is driven into the ground to restrain uncompacted soil near the surface when drilling a caisson is the ___.

_______________ **5.** The pile ___ is the large upper portion of a pile.

_______________ **6.** A(n) ___ is a cast-in-place pile formed by drilling a hole, removing the earth, inserting reinforcement, and filling the hole with concrete.

______________________ **7.** A pile ___ is a metal cone placed over the pile tip to protect the pile from damage while the pile is being driven.

______________________ **8.** A(n) ___ table provides the correct spacing for studs and holds the studs in position while sheathing is nailed to them when constructing large panel forms.

______________________ **9.** A maximum control joint spacing of ___′ is required between joints on a wall with no openings.

______________________ **10.** A(n) ___ wall is a load-bearing wall built below ground level that carries the load transferred to it by load-bearing walls and non-load-bearing walls.

______________________ **11.** The height of a column is typically ___ times its largest dimension.

______________________ **12.** A(n) ___ wall is a non-load-bearing wall that encloses a building and that is supported and anchored to the structural frame.

______________________ **13.** A(n) ___ is a large horizontal structural member that supports beams and other bending loads.

______________________ **14.** Any load that is not permanently applied to a structure is a(n) ___ load.

______________________ **15.** A(n) ___ footing is a foundation footing for a pier or column.

______________________ **16.** Slabs for residential construction are commonly ___″ thick and may or may not be reinforced.

______________________ **17.** ___ is the upward force on a structure resulting from pressure exerted on the structure.

______________________ **18.** A(n) ___ is a groove formed into fresh concrete that interlocks concrete structures placed at different times.

______________________ **19.** A(n) ___ foundation is a spread footing containing a formed keyway and an independently placed foundation wall erected on it.

______________________ **20.** A(n) ___ foundation is used in residential structures consisting of a short wall built on a spread footing that provides space between the bottom of the floor and the ground.

______________________ **21.** A footing that supports more than one column is a(n) ___ footing.

______________________ **22.** A(n) ___ foundation is a footing and wall cast as a single structure.

______________________ **23.** A(n) ___ foundation consists of an independently placed wall above a spread footing that extends on both sides of the wall.

______________________ **24.** A(n) ___ foundation is commonly used on steeply sloped lots.

______________________ **25.** A(n) ___ is a reinforced concrete beam placed at ground level and used to tie walls or column footings together.

______________________ **26.** A(n) ___ foundation has a footing on only one side of the foundation wall.

______________________ **27.** The front ___ is the distance from the building line to the front property line.

______________________ **28.** Insulation under slab-on-grade foundations should be 1″ to 2″ thick and extend horizontally 24″ under the floor slab or vertically ___″ below grade.

______________________ **29.** ___ provide entrance and exit ramps for highways, as well as a way for surface streets to cross a highway.

______________________ **30.** A(n) ___ is a vertical support that provides load-bearing support in the ground and functions similarly to a column.

______________ **31.** A highway ___ is a concrete surface that supports a traffic load.

______________ **32.** Piers beneath grade beams should have a minimum diameter of ___″ and may be flared at their base to cover a wider soil area.

______________ **33.** A(n) ___ pier cap is a wide pier cap that is erected on a single round, rectangular, or square pier.

______________ **34.** Total ___ is the horizontal length of a stairway measured from the foot of the stairway to the point where the stairway ends.

______________ **35.** A(n) ___ is a watertight enclosure used to allow construction or repairs to be performed below the surface of water.

______________ **36.** The end structure that supports beams, girders, the deck of a bridge or arch, and the edges of a bridge is the ___.

______________ **37.** ___ concrete is decorative concrete that will be permanently exposed to view and that requires specially selected concrete materials, forming, placing, and finishing to obtain the desired appearance.

______________ **38.** A(n) ___ is a low wall formed along the edges of an overpass or bridge deck.

______________ **39.** A(n) ___ joint is a formed connection made between precast structures.

______________ **40.** ___ concrete is concrete cast into forms and allowed to set prior to placement in a final location.

______________ **41.** ___ is a method of prestressing reinforced concrete in which tendons are tensioned after the concrete has hardened.

______________ **42.** A(n) ___ is a system of forms and supports for producing concrete members.

______________ **43.** ___ is a process that layers multiple precast members onto each other.

______________ **44.** ___ concrete is precast concrete in which internal stresses are introduced to such a degree that tensile stresses resulting from service loads are counteracted to a desired degree.

______________ **45.** ___ construction is a method of construction in which members are cast horizontally at a location adjacent to their eventual position.

______________ **46.** ___ systems consist of an insert embedded in a wall panel and a lifting unit, which are released from a remote location.

Short Answer

1. Summarize the procedure for drilling a caisson.

____

____

____

____

2. List five advantages of precast concrete over cast-in-place concrete.

____

____

____

____

____

____

3. Summarize the procedure for erecting batterboards and setting building lines for a foundation.

____

____

____

____

____

____

Matching

Piles

__________ **1.** Pile cutoff

__________ **2.** Tip

__________ **3.** Driving head

__________ **4.** Pile head

__________ **5.** Pile foot

__________ **6.** Pile shoe

__________ **7.** Pile butt

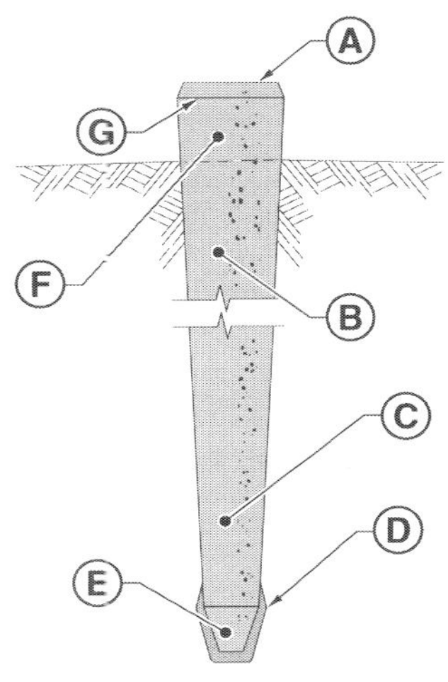

Waterstops

_______________ 1. Cellular

_______________ 2. Single piece with keyway

_______________ 3. Split-fin

_______________ 4. Labyrinth

_______________ 5. Single piece

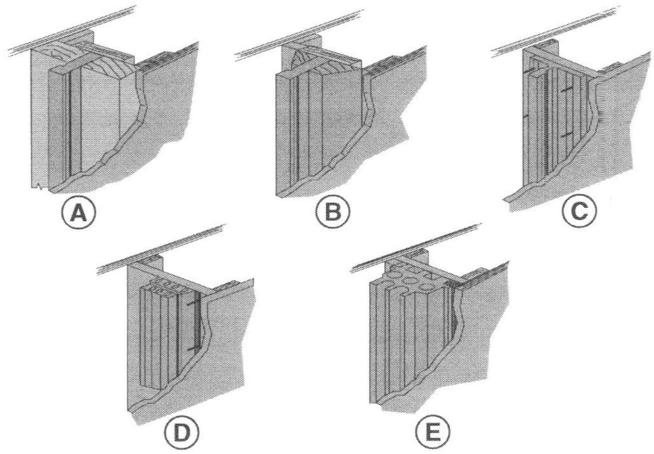

Column Forms

_______________ 1. Brace

_______________ 2. Column clamp

_______________ 3. Cleanout door

_______________ 4. Chamfer strip

_______________ 5. Column form

_______________ 6. Template

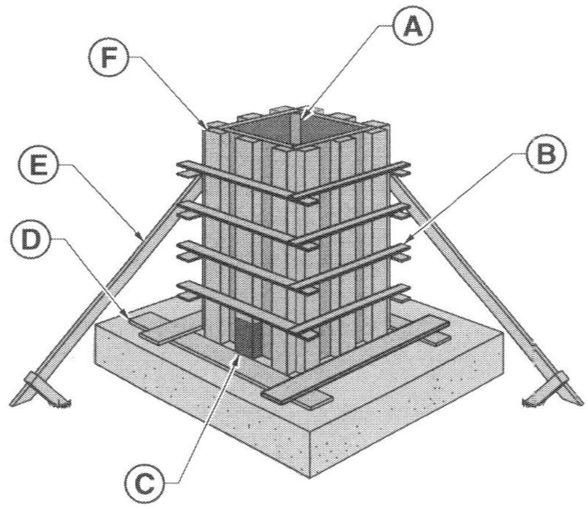

Girders and Beams

_______________ 1. Floor slab

_______________ 2. Beam

_______________ 3. Column

_______________ 4. Girder

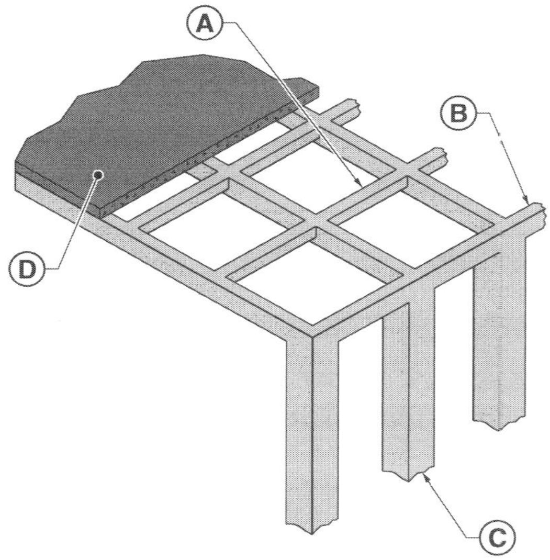

Substructure

______________________ **1.** Abutment

______________________ **2.** Bridge deck

______________________ **3.** Pier

______________________ **4.** Pier cap

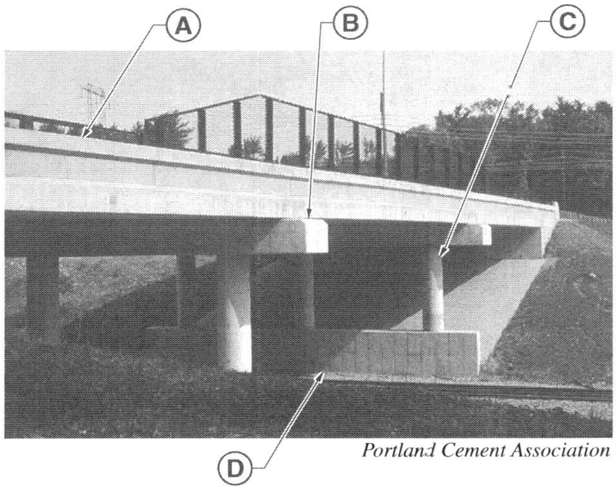

Portland Cement Association

Tilt-Up Hardware

______________________ **1.** Plate anchor

______________________ **2.** Wall brace insert

______________________ **3.** Spread anchor

______________________ **4.** Lifting coil insert

______________________ **5.** Edge lifting coil insert

______________________ **6.** Lift plate

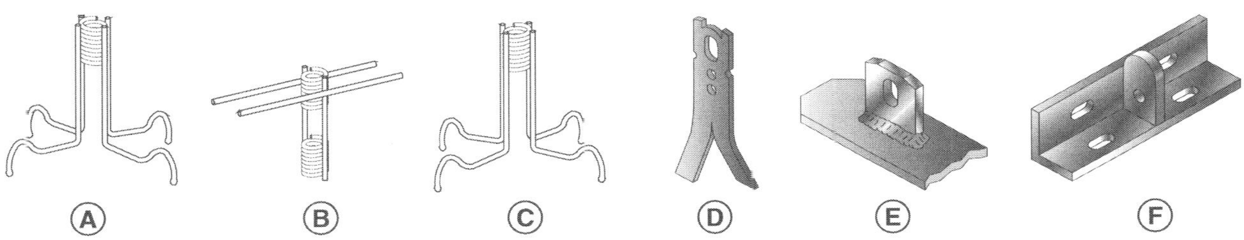

Lifting Units

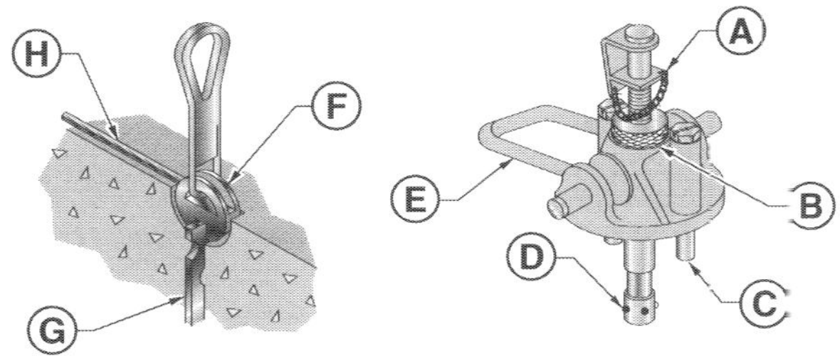

______ **1.** Clutch bar

______ **2.** Safety stop key

______ **3.** T-bar anchor

______ **4.** Spring-loaded plunger

______ **5.** Shackle

______ **6.** Clutch ring

______ **7.** Adjustment mechanism

______ **8.** Encasement ball

Activities

Stairway Layout

Local building codes cite maximum riser height and minimum tread width for stairways. While building codes vary, most codes identify a maximum riser height of 7¾″. Stair riser height is determined by dividing the total rise of the stairway by the number of risers. For example, a stairway with a total rise of 50¾″ containing six stairs has a riser height of 8¼″ (50.75″ ÷ 7 = 7.25″).

Determine the riser height of the following stairways:

______ **1.** Total rise = 70″; 10 stairs

______ **2.** Total rise = 62″; 8 stairs

______ **3.** Total rise = 52½″; 7 stairs

______ **4.** Total rise = 45¾″; 6 stairs

______ **5.** Total rise = 28½″; 4 stairs

Projects

1. Visit a construction site where tilt-up construction is being utilized. Note the equipment used to position the precast panels and how workers communicate with equipment operators. Refer to the textbook appendix to gain an understanding of the meaning of the hand signals used by the construction crew.

Concrete Principles

Chapter 6

Concrete Consolidation

Name ______________________________ **Date** ____________________

True-False

T F **1.** Jitterbugging is the process of consolidating fresh concrete by repeated blows with a jitterbug.

T F **2.** Vibrational frequency is expressed in hertz.

T F **3.** Tamping rods are primarily used for consolidating concrete in wall forms when low-slump concrete is used.

T F **4.** Concrete vibrated at a high frequency yields higher compressive strength than concrete vibrated at a low frequency under similar conditions.

T F **5.** Air pressure used to power a pneumatic vibrator should be allowed to fluctuate to ensure consistent vibrational frequency.

T F **6.** The maximum acceptable vibrational frequency is 12,000 vpm to 14,000 vpm.

T F **7.** The surface of overvibrated concrete is dry and consists of a layer of mortar containing a large amount of coarse aggregate.

T F **8.** Form vibrators are not typically used with wood forms due to the dampening effect of wood on vibration waves.

T F **9.** Amplitude increases as frequency increases since the vibrator head does not have time to travel as far as it can at a higher frequency.

T F **10.** A 180 Hz motor is commonly powered by a special generator.

T F **11.** Centrifugal force is measured in pounds of force.

T F **12.** When operating an internal vibrator, the vibrator should be withdrawn at a rate of approximately 1″ per second.

T F **13.** Vibratory truss screeds generate 3000 vpm to 6000 vpm.

T F **14.** Normal-frequency external vibrators generate a maximum of 1700 lb of centrifugal force.

T F **15.** Rapid withdrawal of a vibrator head from concrete is the primary cause of improper consolidation.

T F **16.** Transmission channels are positioned parallel to the rotational axis of the vibrator.

T F **17.** When an internal vibrator is immersed into concrete, the rpm of the motor typically increases.

T F **18.** Revibrating close to joints between lifts destroys the bond by reducing entrapped air along the joint.

T F **19.** When blending two lifts of concrete, the vibrator should be moved up and down rapidly for 4 sec to 6 sec where the two lifts meet.

Multiple Choice

______ **1.** Depending on lift, the concrete mixture, and the vibrator size or power, vibration can take from ___ in one location.
A. 5 sec to 15 sec
B. 10 sec to 20 sec
C. 15 sec to 30 sec
D. 45 sec to 60 sec

______ **2.** Consolidation is achieved through ___.
A. tamping
B. vibration
C. rodding
D. all of the above

______ **3.** Natural resonant ___ frequency is the frequency at which concrete particles vibrate, rotate, and then consolidate under their own weight.
A. vibrational
B. aggregate
C. amplitude
D. none of the above

______ **4.** The maximum displacement from the neutral axis about which a vibrator moves is the ___.
A. vibrational frequency
B. natural resonant aggregate frequency
C. amplitude
D. torque

______ **5.** Vibration equipment is selected based on ___.
A. formwork design
B. mixture design
C. job size and condition
D. all of the above

______ **6.** A(n) ___ vibrator generates and transmits vibration waves from the exterior to the interior of concrete.
A. internal
B. external
C. immersion
D. none of the above

__________ **7.** A factor that determines the field of action of a vibrator is the ___.

A. slump of the concrete
B. vibrator rpm and compression force
C. size of aggregate in the mixture
D. all of the above

__________ **8.** Each ___% of entrapped air can reduce compressive strength by 3% to 5% at a constant water-cement ratio.

A. 1
B. 3
C. 5
D. 7

__________ **9.** A(n) ___ is a compact test instrument used to measure vibrational frequency.

A. stroboscope
B. amplitude meter
C. sirometer
D. all of the above

__________ **10.** External vibrators are used for ___-slump concrete mixtures or for jobs where spacing of steel reinforcement or form dimensions does not allow for immersion of an internal vibrator.

A. no
B. low
C. high
D. none of the above

__________ **11.** When operating an internal vibrator, distribution of force within concrete depends on the ___ of the vibrating head.

A. surface area
B. size
C. shape
D. all of the above

__________ **12.** High-frequency, low-amplitude internal vibrators are commonly used for general consolidation work including ___.

A. footings
B. slabs
C. walls
D. all of the above

__________ **13.** A(n) ___ vibrator is immersed directly into concrete and transmits vibration waves to consolidate a concrete mixture.

A. external
B. surface
C. internal
D. form

______________ **14.** The force of compression impulses generated by a vibrator is determined by ___.

A. eccentric weight mass
B. speed of the vibrator
C. radius from the center of rotation to the center of gravity of the eccentric weight
D. all of the above

______________ **15.** ___ is the number of times a vibrator head moves from side to side in one minute.

A. Vibrational frequency
B. Aggregate frequency
C. Amplitude
D. Torque

______________ **16.** An internal vibrator generating ___ should be used for concrete with fine aggregate.

A. 3000 vpm to 6000 vpm
B. 6000 vpm to 9000 vpm
C. 9000 vpm to 12,000 vpm
D. 12,000 vpm to 15,000 vpm

______________ **17.** A(n) ___ vibrator is used for consolidating concrete for precast units.

A. form
B. table
C. internal
D. immersion

______________ **18.** Intermediate-frequency external vibrators are used for ___.

A. vibrating trestles
B. vibrating tables
C. ganged panel forms
D. all of the above

______________ **19.** ___ vibrators generate a maximum of 1500 lb of centrifugal force.

A. Normal-frequency
B. Intermediate-frequency
C. High-frequency
D. none of the above

______________ **20.** A form vibrator vibrates concrete indirectly at a frequency of ___.

A. 3000 vpm to 6000 vpm
B. 6000 vpm to 12,000 vpm
C. 12,000 vpm to 15,000 vpm
D. 15,000 vpm to 18,000 vpm

______________ **21.** The size of external vibrator required for a concrete structure can be approximately calculated as 1 lb of centrifugal force equaling ___ of mass.

A. 1 lb to 2 lb
B. 2 lb to 5 lb
C. 5 lb to 7 lb
D. 7 lb to 10 lb

_______________ **22.** A vibrator generating ___ should be used for concrete with primarily medium aggregate.
- A. 3000 vpm to 6000 vpm
- B. 6000 vpm to 9000 vpm
- C. 9000 vpm to 12,000 vpm
- D. 12,000 vpm to 15,000 vpm

_______________ **23.** A vibrator generating ___ should be used for concrete with primarily coarse aggregate.
- A. 3000 vpm to 6000 vpm
- B. 6000 vpm to 9000 vpm
- C. 9000 vpm to 12,000 vpm
- D. 12,000 vpm to 15,000 vpm

_______________ **24.** When using an internal vibrator, an operator should maintain a ___″ space between the vibrator head and the form.
- A. 1
- B. 2
- C. 4
- D. 6

_______________ **25.** When using an internal vibrator, the distance between insertion points should be approximately ___ times the radius of the field of action.
- A. 1½
- B. 2
- C. 2½
- D. 3

_______________ **26.** External vibrators are attached to one side of wall formwork for walls up to ___″ thick.
- A. 4
- B. 6
- C. 10
- D. 12

_______________ **27.** ___ is the process of vibrating concrete after the concrete has been placed and initially consolidated but before the initial setting of the concrete.
- A. Undervibration
- B. Overvibration
- C. Revibration
- D. none of the above

_______________ **28.** An advantage of using normal-frequency vibrators is ___.
- A. an increased field of action
- B. an ability to handle products with high mass
- C. low cost
- D. all of the above

_______________ **29.** When vibrating a lift of fresh concrete with an internal vibrator, immerse the vibrator head through the new lift and approximately ___ into the previous lift.

A. 5% to 10%
B. 15% to 20%
C. 25% to 30%
D. 35% to 40%

_______________ **30.** A vibratory truss screed should be approximately ___ longer than the span of the area to be screeded.

A. 1″ to 3″
B. 3″ to 5″
C. 5″ to 8″
D. 6″ to 9″

_______________ **31.** Revibration ___.

A. increases compressive strength by 20% to 25%
B. releases additional bleedwater trapped under steel reinforcement
C. improves the bonding capability of the concrete
D. all of the above

_______________ **32.** When using a vibratory truss screed, concrete should be spread approximately ___″ above the bottom of the leading bar.

A. 1
B. 2
C. 4
D. 6

_______________ **33.** A ___ is a braced frame that is used to support the vibrating table, pallet, form, and concrete.

A. casting bed
B. vibrating trestle
C. vibrating truss
D. none of the above

_______________ **34.** ___ is the separation of the concrete ingredients into separate horizontal layers, with the lighter ingredients close to the top.

A. Segregation
B. Stratification
C. Reverberation
D. none of the above

Completion

_______________ **1.** Acceleration is measured in ___.

_______________ **2.** ___ is the process of consolidating fresh concrete using a tamping rod.

_______________ **3.** ___ is the excessive use of consolidation equipment during placement of fresh concrete.

_______________ **4.** ___ is the process of creating a close arrangement of solid particles in fresh concrete during placement by reducing the space between the particles.

_______________ **5.** The field of ___ is the area of concrete affected by the compression impulses of the vibrator.

_______________ **6.** Although each particle size has its own natural resonant aggregate frequency, a normal concrete mixture, which includes fine and coarse aggregate, water, and cement, has a natural frequency of approximately ___ vpm.

_______________ **7.** ___ is the force produced by the rotating eccentric weight within the vibrator and is measured in pounds of force.

_______________ **8.** ___ is a series of rapid compression impulses generated by an eccentric weight rotating at a high rpm.

_______________ **9.** The electrical frequency of a motor is measured in cycles per second, or ___.

_______________ **10.** A vibratory truss screed is used to strike off and consolidate low-slump concrete sections in slabs up to ___′ wide.

_______________ **11.** A(n) ___ vibrator is attached to selected positions on form exteriors.

_______________ **12.** When using an internal vibrator, avoid vibrating too quickly and in a(n) ___ pattern.

_______________ **13.** When using a vibratory truss screed, one worker should be positioned every ___′ of screed length to spread the concrete evenly.

_______________ **14.** A(n) ___ vibrator is a vibrator that employs a portable horizontal platform upon which a vibrating element is mounted.

_______________ **15.** Intermediate-frequency vibrators generate a maximum of ___ lb of centrifugal force.

_______________ **16.** ___ is the total weight of the concrete formwork and concrete mixture, which is used to calculate vibrator size.

_______________ **17.** ___ is the lack of complete consolidation during placement of fresh concrete.

_______________ **18.** A(n) ___ channel is a section of metal channel attached to the formwork that transmits vibration waves and prevents damage to the formwork.

Short Answer

1. List five advantages of external vibrators.

____

____

____

____

____

2. List five benefits of concrete consolidation by vibration.

3. List the three factors that determine the field of action.

4. List four considerations when selecting the type of internal vibrator to be used for a job.

5. List three general guidelines for using an internal vibrator.

6. List three general guidelines for using external vibrators on precast forms.

7. Summarize the procedure for aligning a vibratory truss screed.

Matching

Motor-in-Head Vibrator

______________ **1.** Armature core

______________ **2.** Nose

______________ **3.** Eccentric weight

______________ **4.** Collar

______________ **5.** Bearing

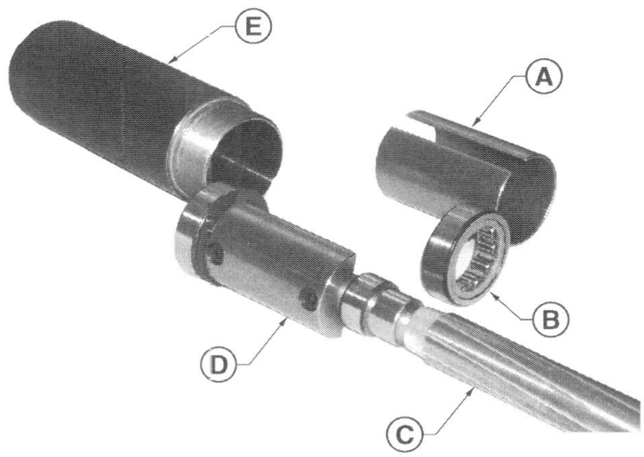

Eccentric Weight Configurations

______________ **1.** Weighted disc

______________ **2.** Rotor shaft

______________ **3.** Square key

______________ **4.** Nut

______________ **5.** Taper pin

______________ **6.** Stationary weight

______________ **7.** Adjustable weight

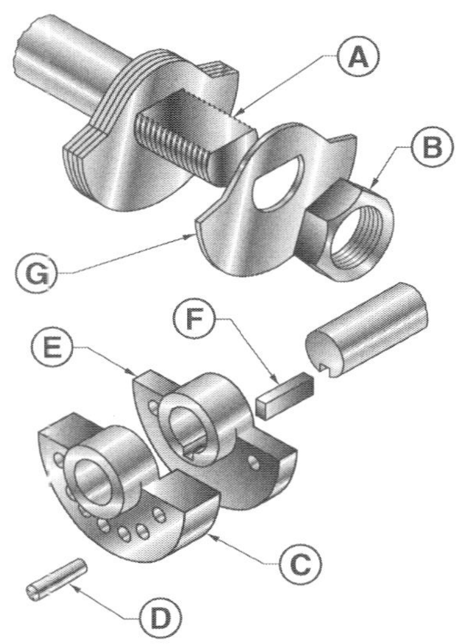

Activities

Vibration Rates

Small vibrators can commonly vibrate 5 cu yd/hr to 10 cu yd/hr, and heavy-duty vibrators can commonly vibrate 50 cu yd/hr. The volume of concrete is calculated by multiplying the thickness (in ft) by the width (in ft) by the length (in ft) and dividing by 27. The number of hours needed to vibrate a given volume of concrete is calculated as follows:

$$T = V \div r$$

where

T = time (in hr)
V = volume (in cu yd)
r = rate (in cu yd/hr)

For example, determine the amount of time required to vibrate a concrete slab measuring 6″ (0.5′) thick by 50′ wide by 75′ long when the vibration rate is 10 cu yd/hr.

$$V = \frac{t \times w \times l}{27}$$

$$V = \frac{0.5 \times 50 \times 75}{27}$$

$$V = 69.5 \text{ cu yd}$$

$$T = V \div r$$
$$T = 69.5 \div 10$$
$$T = \mathbf{6.9\ hr}$$

Determine the amount of time required to vibrate the following volumes of concrete:

______________________ **1.** 30 cu yd of concrete vibrated at 7.5 cu yd/hr

______________________ **2.** 45 cu yd of concrete vibrated at 5 cu yd/hr

______________________ **3.** 100 cu yd of concrete vibrated at 30 cu yd/hr

______________________ **4.** 125 cu yd of concrete vibrated at 10 cu yd/hr

______________________ **5.** 125 cu yd of concrete vibrated at 35 cu yd/hr

Projects

1. Obtain product catalogs from various manufacturers of internal and external vibrators. Compare and contrast the features, speed, weight, and size of the vibrators and prepare a comparison table of this information.

2. Contact precast concrete suppliers to learn more about their operations. Identify the most commonly manufactured precast members and investigate the placement of vibrators on the formwork.

Concrete Principles

Chapter 7

Concrete Finishing

Name ______________________ **Date** ______________

True-False

T F **1.** Large piles of concrete are distributed in forms by dragging the rake across the concrete with the teeth side up.

T F **2.** Jitterbugging produces an evenly consolidated slab.

T F **3.** Hand floating is a floating operation used to fill in low spots and level down ridges left by the strikeoff operation.

T F **4.** Bull floating is performed before screeding.

T F **5.** Hand floating prepares the surface for finish troweling or brooming.

T F **6.** Power floating uses an engine-driven power trowel for floating.

T F **7.** A steel trowel is used for the final finishing operation and is commonly used in conjunction with a hand float.

T F **8.** During steel troweling, the tilt of the trowel blade is slightly decreased and less downward pressure is applied for each subsequent troweling pass.

T F **9.** Finishing with a walk-behind power trowel can begin earlier than finishing with a ride-on power trowel.

T F **10.** Non-overlapping power trowels can accommodate any type of trowel blade and pans.

T F **11.** A combination blade reduces the weight per square inch of the power trowel during floating when concrete is still plastic.

T F **12.** When concrete is too wet, a ride-on power trowel with pans is difficult to steer.

T F **13.** A sidewalk edger is an edging tool used to finish slabs and is available in common radius sizes of ⅛″, ¼″, ½″, and ¾″.

T F **14.** When placing concrete for a large slab, concrete placement should begin along one edge of the formwork and should work toward the middle in long strips.

T F **15.** Jitterbugging is an acceptable method for consolidating sidewalks and small slabs.

T F **16.** Floating seals the concrete surface, preventing bleedwater from surfacing.

T F **17.** When pushing a bull float across a slab, the back edge of the bull float should be lifted slightly to avoid digging into the concrete.

T F **18.** Edging is easier if concrete is separated from the forms after the floating stage before concrete bonds to the forms.

T F **19.** Concrete should be wetted during finishing to bring a layer of cement paste to the surface.

T F **20.** An overlapping ride-on power trowel can use any type of blades, but it cannot use pans.

T F **21.** High overall flatness on a large slab can be best achieved with non-overlapping power trowels.

T F **22.** When using pans with a power trowel, the blades should be adjusted to flat or very slight pitch to remove the slack in the pitching mechanism to prevent chatter.

T F **23.** When concrete is being placed and is near grade level, the flat edge of the rake is used to pre-level the surface.

T F **24.** Consolidation by vibration converts stiff concrete into a fluid state, which makes concrete easier to work during finishing.

T F **25.** On large-scale jobs, a vibratory truss screed is used for consolidation and screeding.

T F **26.** The second bull float operation is performed perpendicular to the first bull float operation.

T F **27.** Steel trowels are commonly 7½″ to 2′ long.

T F **28.** A trowel should make a ringing, scraping sound as it is swept across the surface of the concrete on its final pass.

T F **29.** New power trowel blades that are not wear-rounded should be operated flat and rotated at a low speed.

T F **30.** Power trowel pans are used for power floating and are rotated at a relatively slow speed.

T F **31.** A slab that has not been floated may be troweled even though some coarse aggregate at the surface may be dislodged by the edge of the trowel blade.

T F **32.** Edging should not begin until a person standing on the slab does not make an indentation of more than ¼″.

T F **33.** Water moves more quickly through air-entrained concrete, increasing the bleedwater at the surface.

T F **34.** A margin trowel is used to patch small areas.

T F **35.** When using burlap sheets to cure concrete, wet the burlap sheets thoroughly and allow them to dry completely before soaking them again with water.

T F **36.** Concrete sets faster in hot, dry, or windy weather than in moderate conditions with no wind.

T F **37.** A minimum of two stamp pads should be used to properly align a pattern.

T F **38.** A more intense shade of color is possible with pigment admixtures than with face pigments.

T F **39.** Additional coats of concrete stain will not affect the intensity of the surface color.

T F **40.** A margin trowel is used for mixing, scooping, and finishing patch material.

T F **41.** The desired finish of a concrete surface determines the number of times the surface is to be troweled.

T F **42.** Finish blades apply more pressure to concrete than float blades, which increases the density of concrete as it hardens.

T F **43.** Sidewalk edgers are typically larger than highway edgers and are attached to a long handle.

T F **44.** During a period of freezing, the bubbles created with an air-entraining admixture act as a holding chamber into which the freezing water in the concrete can expand without developing pressures great enough to crack the concrete.

T F **45.** A pointing trowel is used to apply patch material and to finish areas that are inaccessible to a standard finishing trowel.

T F **46.** Power trowel blades are made of aluminum.

T F **47.** Strong, high-quality coloring pigments should be used since they require less pigment to attain the desired color.

T F **48.** Chromium oxide is used to produce blue-colored concrete.

T F **49.** Dry shakes should never be distributed onto concrete that has bleedwater present or has not been floated.

T F **50.** Combination blades are often used for an entire job, floating at a high speed and troweling at a low speed.

T F **51.** Float blades attach directly to trowel arms.

T F **52.** Non-air-entrained concrete is used extensively in exposed concrete structures such as bridges, buildings, concrete pavements, and curbs.

T F **53.** Aluminum or magnesium floats are used if floating is done by hand on air-entrained concrete.

T F **54.** Concrete gains no strength when frozen.

T F **55.** When using ponding as a curing technique, the pond can be drained at the end of the curing period and the slab can be used immediately.

T F **56.** White-pigmented liquid-membrane compounds reflect heat rays and help maintain a steady concrete temperature.

T F **57.** Building paper is one type of waterproof paper that can be used for curing concrete.

T F **58.** A job site must be well-ventilated when applying a liquid-membrane curing compound.

T F **59.** A broom finish is a non-slip finish that uses a specially designed broom pushed and pulled over concrete to achieve a variety of patterns.

T F **60.** The rock salt and travertine finishing methods are not recommended for surfaces that are exposed to freeze/thaw cycles.

T F **61.** Concrete that freezes within the first 24 hr after placement is likely to be permanently damaged.

T F **62.** Colored concrete must be vibrated thoroughly to prevent bug holes and rock pockets from occurring.

T F **63.** Calcium chloride or other chlorides should not be used in colored concrete.

T F **64.** Concrete cures more quickly at cold temperatures.

T F **65.** Spraying is a curing process that produces a steady, fine spray of water, increasing hydration and assisting proper curing.

T F **66.** Liquid-membrane compounds should be applied in two light passes made at right angles.

T F **67.** Black liquid-membrane compounds provide an effective moisture barrier and are recommended whenever color is unimportant.

T F **68.** The most appropriate time to place concrete in hot, dry, or windy weather is early in the morning or late in the evening.

T F **69.** Concrete that has frozen cannot be salvaged even if proper curing is resumed after the concrete thaws.

T F **70.** Steel trowels produce the coarse textures necessary for steep slopes or in areas of heavy traffic.

T F **71.** Aggregate should be dislodged when creating an exposed aggregate finish

T F **72.** Placing and finishing procedures for colored concrete are the same as for regular concrete.

T F **73.** An exposed aggregate finish is a texturing pattern where the surface layer of cement has been removed to expose aggregate material embedded in the concrete.

T F **74.** When floating a slab that has recently had dry shakes applied to it, the edges of the slab should be floated first.

T F **75.** Dry shakes provide greater uniformity of color if applied in a single, heavy coat.

T F **76.** Column forms should be constructed so that the sides can be removed without disturbing the adjoining beam or girder forms.

T F **77.** Liquid surface treatments are not intended to provide additional wear-resistance in new, well-designed, well-constructed, and cured floors.

Multiple Choice

______________ **1.** Finishing procedures include ___ of concrete.

A. placement
B. curing
C. finishing
D. all of the above

____________________ **2.** ___ is a leveling process of striking off a concrete surface using guides and a straightedge or a vibratory truss screed.

A. Straightedging
B. Jitterbugging
C. Screeding
D. Tamping

____________________ **3.** Bull floats are available in sizes from ___ in length.

A. 6″ to 2′
B. 3′ to 4′
C. 3′ to 10′
D. 4′ to 12′

____________________ **4.** Hand floats are available in lengths from ___.

A. 3″ to 6″
B. 6″ to 12″
C. 12″ to 24″
D. 24″ to 36″

____________________ **5.** A common steel trowel size is ___ wide and 7½″ to 2′ long.

A. 3″ to 5″
B. 5″ to 10″
C. 8″ to 14″
D. 10″ to 12″

____________________ **6.** A ___ trowel is used to patch small areas and is usually 1″ to 2″ wide and 5″ to 8″ long.

A. steel
B. fresno
C. pointing
D. margin

____________________ **7.** Concrete reaches about ___% of final strength in the first seven days of curing.

A. 40
B. 50
C. 60
D. 70

____________________ **8.** A common non-slip finish is ___.

A. broom finish
B. wire combing
C. floated swirl finish
D. all of the above

____________________ **9.** Medium textures, where some slip-resistance is necessary, are produced by ___ floats.

A. canvas resin
B. magnesium
C. aluminum
D. all of the above

______________ **10.** ___ percent of entrapped air can reduce the compressive strength of concrete by 5% to 6%.

A. One
B. Two
C. Three
D. none of the above

______________ **11.** A footprint indentation on the concrete surface should not be deeper than about ___″ before using float blades.

A. ⅛
B. ¼
C. ⅜
D. ¾

______________ **12.** A float blade is used in floating to ___.

A. embed coarse aggregate
B. compact the surface
C. open up the concrete surface
D. all of the above

______________ **13.** ___ floats should only be used on non-air-entrained concrete.

A. Magnesium
B. Wood
C. Stainless steel
D. none of the above

______________ **14.** When producing an exposed aggregate finish, uniformly-sized aggregate, usually ___″ or larger, is evenly distributed on the concrete surface immediately after darby floating.

A. ⅛
B. ¼
C. ⅜
D. ¾

______________ **15.** When producing an exposed aggregate finish, a(n) ___ can be used on the concrete surface if exposing of the aggregate is delayed.

A. accelerator
B. air-entraining admixture
C. retarder
D. none of the above

______________ **16.** When coloring concrete using a coloring pigment, concrete slump should be between ___ to ensure proper mixing of color without adding excessive water to the mixture.

A. 1″ and 3″
B. 2″ and 4″
C. 3″ and 5″
D. 4″ and 6″

_______________ **17.** Vibrating concrete ___.
A. provides greater density
B. improves the bond between concrete and steel reinforcement
C. increases the tensile strength
D. all of the above

_______________ **18.** Finishing procedures include ___.
A. consolidating
B. troweling
C. raking and spreading
D. all of the above

_______________ **19.** The timing of floating is affected by ___.
A. humidity
B. slump
C. concrete temperature
D. all of the above

_______________ **20.** During finishing stages, pans may be used for ___.
A. opening a surface that dried too soon
B. preparing for a final broom finish
C. correcting discoloration caused by rain that has fallen on the surface
D. all of the above

_______________ **21.** Improper curing may result in ___.
A. cracking
B. spalling
C. curling
D. all of the above

_______________ **22.** The depth of a concrete surface texture is determined by ___.
A. bristle stiffness
B. pressure
C. surface workability
D. all of the above

_______________ **23.** When screeding a surface, approximately ___″ of excess concrete should be in front of the straightedge so material is available for leveling the surface.
A. 1
B. 2
C. 3
D. 5

_______________ **24.** Slump loss affects the ___ of the concrete.
A. workability
B. durability
C. consistency
D. all of the above

______________ **25.** A ___ high windrow created between the pans provides the proper amount of mortar to minimize restraightening efforts and is the signal to begin pan floating.
A. ⅛″ to ¼″
B. ¼″ to ½″
C. ½″ to ¾″
D. ¾″ to 1″

______________ **26.** ___ of a concrete surface may result if concrete is finish-troweled while still plastic.
A. Dusting
B. Blistering
C. Scaling
D. all of the above

______________ **27.** Floats are made of ___.
A. wood
B. magnesium
C. aluminum
D. all of the above

______________ **28.** Finish troweling helps ___ the concrete surface, making it smooth and durable.
A. harden
B. compact
C. compress
D. all of the above

______________ **29.** Darby floats are available in sizes from ___ in length and have from 1 to 3 handles.
A. 6″ to 12″
B. 12″ to 18″
C. 18″ to 24″
D. 24″ to 48″

______________ **30.** To maintain a uniform-colored concrete, ___.
A. use the same proportion of materials in each batch
B. mix each batch for the same length of time
C. buy pigment for the whole job at one time
D. all of the above

______________ **31.** Surface hardeners ___.
A. form crystalline deposits
B. penetrate concrete
C. act as a binder to make the floor harder
D. all of the above

______________ **32.** Concrete used for an exposed aggregate finish should have a maximum ___″ slump.
A. 2
B. 3
C. 4
D. 6

_______________ **33.** Full concrete strength is reached after approximately ___ days of proper curing.
A. 7
B. 14
C. 21
D. 28

_______________ **34.** Slabs from ___ thick require an air-driven vibratory truss screed or an internal vibrator used on a horizontal plane.
A. 3″ to 6″
B. 6″ to 9″
C. 9″ to 12″
D. 12″ to 16″

_______________ **35.** A walk-behind power trowel ___.
A. has one rotor with three or four blades
B. has a guard ring diameter from 24″ to 50″
C. is powered by a gasoline engine
D. all of the above

_______________ **36.** Floating ___.
A. embeds large pieces of aggregate beneath the surface
B. compacts surface concrete
C. brings sufficient mortar to the surface
D. all of the above

_______________ **37.** A common manual floating technique is ___ floating.
A. darby
B. bull
C. hand
D. all of the above

_______________ **38.** Pointing trowels are made of ___ and are about 4″ to 7″ long.
A. aluminum
B. wood
C. steel
D. magnesium

_______________ **39.** A ride-on power trowel ___.
A. is powered by one or two gasoline engines
B. has two rotors, each with four blades
C. is commonly available in 36″, 46″, and 60″ guard ring diameters
D. all of the above

_______________ **40.** Power trowel pans ___.
A. are flat on the bottom
B. are made of iron
C. have the outer rim turned up
D. all of the above

______________ **41.** A ride-on power trowel requires a ___ longer wait before floating than a walk-behind power trowel because of the greater weight of the machine.
A. 10 min to 15 min
B. 20 min to 30 min
C. 30 min to 45 min
D. 40 min to 60 min

______________ **42.** A walk-behind trowel larger than ___″ cannot be safely controlled by the operator because of the excessive drag and friction caused by the pans.
A. 24
B. 36
C. 48
D. 60

______________ **43.** Wet burlap curing is a curing process that uses specially treated, coarsely woven ___ to hold moisture in concrete.
A. flax
B. hemp
C. jute
D. all of the above

______________ **44.** When placing concrete at air temperatures below ___°F, fresh concrete must be protected from freezing.
A. 20
B. 30
C. 40
D. 60

______________ **45.** ___ is a chemical reaction between cement and water that bonds molecules and results in hardening of the concrete mixture.
A. Curing
B. Ponding
C. Hydration
D. none of the above

______________ **46.** Black sheets of polyethylene film are used for curing concrete in ___ weather to absorb heat.
A. hot
B. cool
C. normal
D. none of the above

______________ **47.** Before applying magnesium, zinc fluosilicate, or sodium silicate to a concrete floor, the floor should be ___.
A. at least 28 days old
B. air-dried
C. cured thoroughly
D. all of the above

_______________ **48.** A ___ edger is used to shape curved or irregular surfaces and has the lip and radius on the front of the edger.
A. sidewalk
B. curb
C. narrow
D. walking

_______________ **49.** ___ affects the rate of evaporation from concrete.
A. Wind velocity
B. Relative humidity
C. Concrete temperature
D. all of the above

_______________ **50.** A shake-on surface hardener can be applied to a concrete surface to ___.
A. increase wear-resistance
B. reduce dusting
C. increase slip-resistance
D. all of the above

_______________ **51.** A common edger size is 3″ wide and 6″ long with a ___″ radius.
A. ⅛
B. ¼
C. ½
D. ¾

_______________ **52.** Concrete ___ continues to improve as long as favorable conditions are maintained.
A. strength
B. wear-resistance
C. durability
D. all of the above

_______________ **53.** The amount of spray used when curing concrete depends on ___.
A. air temperature
B. concrete temperature
C. wind rate
D. all of the above

_______________ **54.** Green-colored concrete is produced using ___.
A. chromium oxide
B. raw umber
C. carbon black
D. burnt umber

_______________ **55.** A wear-resistant surface hardener is made of ___.
A. trap rock
B. emery
C. granite
D. all of the above

________________ **56.** A curb edger commonly has a ___″ radius.
A. ⅛
B. ¼
C. ½
D. ¾

________________ **57.** Concrete damage is possible if evaporation rates exceed ___ lb/sq ft/hr.
A. 0.01
B. 0.1
C. 1
D. 10

________________ **58.** Mortar must be disposed of if not used within ___ hr(s) of preparation.
A. 1
B. 2
C. 3
D. 4

________________ **59.** The ideal temperature at which to cure concrete is between ___.
A. 35°F and 50°F
B. 40°F and 60°F
C. 55°F and 73°F
D. 75°F to 90°F

________________ **60.** Specially treated burlap is ___.
A. decay-resistant
B. fireproof
C. plastic- or aluminum-coated
D. all of the above

________________ **61.** Ponding ___.
A. accelerates evaporation
B. exposes the concrete
C. warms the concrete surface
D. seals the concrete

________________ **62.** A type of liquid-membrane compound is ___.
A. yellow
B. blue-pigmented
C. pink
D. black

________________ **63.** Fresh concrete is power floated to bring approximately ___ of cement paste to the surface.
A. ⅛″ to ¼″
B. ¼″ to ½″
C. ½″ to ¾″
D. ¾″ to 1″

_______________ **64.** The ___ is the cutting edge of an edger that separates the concrete from the forms.
A. radius
B. lip
C. handle
D. groove

_______________ **65.** Liquid-membrane compounds may contain ___.
A. epoxy
B. wax
C. rubber
D. all of the above

_______________ **66.** Clear sheets of polyethylene film are used for curing in ___ weather.
A. hot
B. cool
C. normal
D. none of the above

_______________ **67.** Waterproof paper is available in widths from ___.
A. 6″ to 12″
B. 12″ to 24″
C. 18″ to 96″
D. 24″ to 108″

_______________ **68.** Liquid-membrane compounds are ___.
A. difficult to apply
B. environmentally safe
C. nonflammable
D. carcinogenic

_______________ **69.** A(n) ___ pattern is a textured finish that creates a rock-like pattern with random shaping of the concrete.
A. travertine
B. flagstone
C. rock salt
D. exposed aggregate

_______________ **70.** A type of barrier used to protect and cure concrete is ___.
A. polyethylene film
B. a liquid-membrane compound
C. waterproof paper
D. all of the above

_______________ **71.** A ___ trowel should not be used to finish colored concrete.
A. plastic
B. regular steel
C. stainless steel
D. all of the above

______________ **72.** White sheets of polyethylene film are used for curing concrete in ___ weather to reflect heat and light.

A. hot
B. cool
C. normal
D. none of the above

______________ **73.** A surface hardener is a chemical solution of ___.

A. gums
B. sodium silicate
C. fluosilicate of magnesium and zinc
D. all of the above

Completion

______________ **1.** Consolidation can be completed using an engine-driven vibratory truss screed on slabs up to ___″ thick.

______________ **2.** When bull floating a slab, overlap consecutive passes by a minimum of ___″ until the slab is completely floated.

______________ **3.** Power floating is performed at a(n) ___ speed to avoid digging into and slinging fresh concrete.

______________ **4.** A(n) ___ is a hand tool with a long handle and a steel grill base used for compacting wet concrete, forcing large aggregate below the surface, and bringing cement paste to the surface for finishing.

______________ **5.** When steel troweling, tilting the blade too much creates a(n) ___ effect on the surface of the concrete.

______________ **6.** ___ ride-on power trowels have blade rotors with overlapping paths.

______________ **7.** A(n) ___ is a 6″ wide power trowel blade with turned-up ends that is operated at a relatively high speed.

______________ **8.** A(n) ___ is a set of two edgers with matching inside and outside radii.

______________ **9.** Power trowels should not be used on air-entrained concrete with an air content of more than ___%.

______________ **10.** ___ is the use of water to cover concrete during the hydration process.

______________ **11.** ___ paper is flexible plastic material that resists moisture.

______________ **12.** ___ is the process of consolidating fresh concrete by repeated blows with a jitterbug.

______________ **13.** A(n) ___ is a hand tool used to screed concrete to a smooth surface.

______________ **14.** ___ is a procedure that levels ridges left by screeding and fills small hollows in the surface of the concrete.

_______________ **15.** Darby floating is a floating operation using a(n) ___ float to fill in low spots and level down ridges left by the strikeoff operation.

_______________ **16.** Bull floating is a floating operation using a(n) ___ float to fill in low spots and level down ridges left by the strikeoff operation.

_______________ **17.** A(n) ___ trowel is a large trowel that is made of aluminum or steel and typically has a 5″ wide blade and is 2′ to 4′ long.

_______________ **18.** A(n) ___ is an 8″ wide power trowel blade with the leading edge and both ends turned up.

_______________ **19.** A(n) ___ trowel is made of steel, is about 4″ to 7″ long, and is used for preparing small areas for patching.

_______________ **20.** A(n) ___ blade is a power trowel blade that is 10″ wide with all four edges turned up.

_______________ **21.** A(n) ___ edger is used for curb and gutter radii.

_______________ **22.** During initial curing, concrete temperature should be maintained at approximately ___°F and concrete should be kept thoroughly moist for a minimum of three days.

_______________ **23.** A liquid-membrane compound is sprayed onto fresh concrete to form a chemical barrier to prevent the loss of ___ from the concrete.

_______________ **24.** A(n) ___ is a coloring agent added to clear compounds to make them visible for application that fades after a few days.

_______________ **25.** A(n) ___ is the distance that one plastic sheet overlays another.

_______________ **26.** Concrete ___ is a method of applying a rough or grooved decorative finish to concrete.

_______________ **27.** Finish blades and combination blades attach to the power trowel ___.

_______________ **28.** The ___ is the curve produced when the blade and lip of an edger meet.

_______________ **29.** Concrete ___ is the process of maintaining proper concrete moisture content and concrete temperature long enough to allow hydration of concrete to occur.

_______________ **30.** ___ is a putty-like adhesive that maintains its elasticity after setting.

_______________ **31.** A(n) ___ treatment is applied after finishing procedures, while concrete is still workable, to improve safety or to give concrete a distinctive or attractive appearance.

_______________ **32.** A(n) ___ is a texturing pattern that is produced by a special hand float to create a fan-like effect.

_______________ **33.** On windy days, ___ should be built to keep water from blowing off the curing site.

_______________ **34.** Approximately 85% of final concrete strength is reached after ___ days of curing.

_______________ **35.** ___ is a curing process that produces a mist-like spray of water to increase hydration and assist proper curing.

_______________ **36.** ___ film is used as a vapor barrier and waterproofing for concrete curing.

______________ **37.** ___ is a texturing procedure that produces a scored surface on concrete.

______________ **38.** A(n) ___ finish is a bumpy surface with high and low spots.

______________ **39.** Concrete ___ is a procedure used to create brick, cobblestone, tile, or other patterns in concrete.

______________ **40.** Ponding for ___ days is the preferred curing method for colored concrete.

______________ **41.** Pigment admixtures can comprise up to ___% of the total volume of concrete.

______________ **42.** A(n) ___ is a powder that is shaken onto and floated into fresh concrete to change the color of concrete.

______________ **43.** A(n) ___-resistant surface hardener is made of silicon carbide or aluminum oxide.

Short Answer

1. Describe the procedure for screeding a slab using a straightedge.

2. Summarize the procedure for hand floating.

3. Describe the procedure for the first steel troweling.

4. List five guidelines for power trowel operation during finish troweling.

5. List six power trowel safety rules.

6. List five hot weather curing considerations.

7. Identify five techniques that can be used to achieve faster curing in cold weather.

8. List the steps required for producing a broom finish.

9. Describe the proper procedure for creating an exposed aggregate finish.

10. List three guidelines for achieving high-quality colored concrete.

Matching

Hand Trowels

____________________ **1.** Fresno

____________________ **2.** Margin

____________________ **3.** Pointing

____________________ **4.** Steel

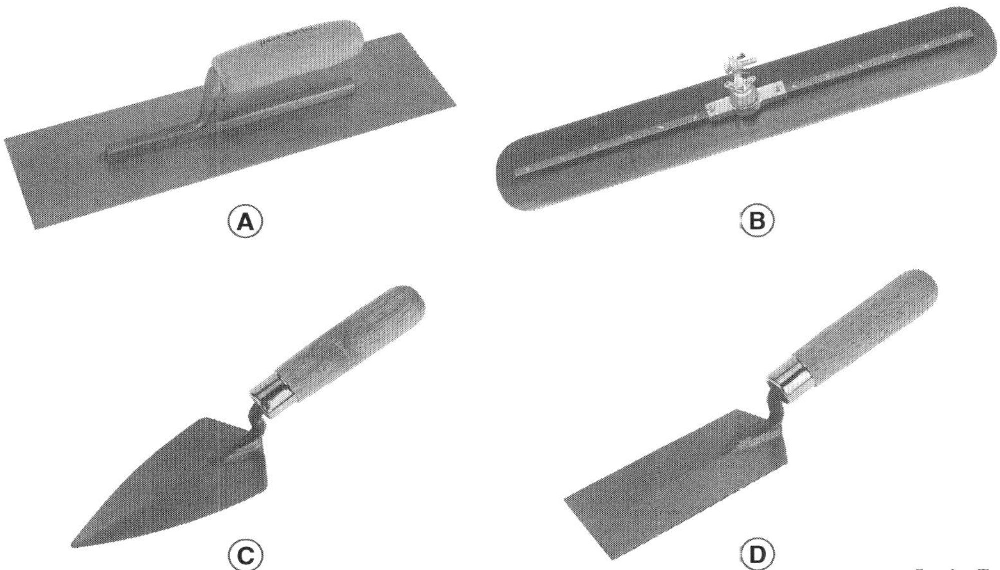

Stanley Tools

Power Trowel Blades

____________________ **1.** Finish

____________________ **2.** Float

____________________ **3.** Combination

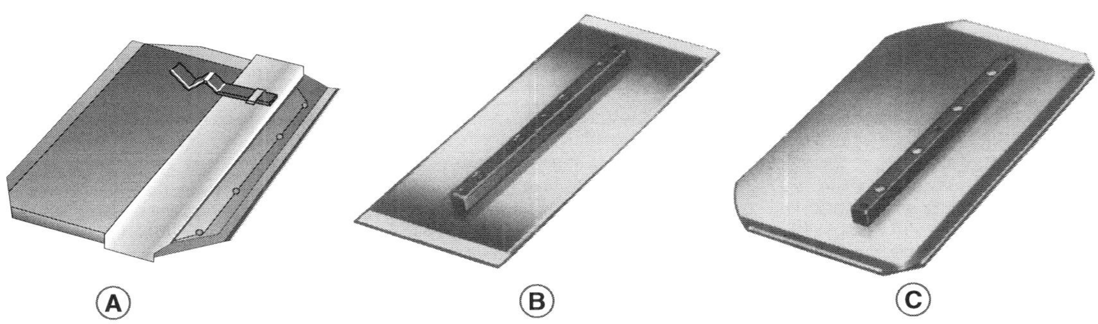

Activities

Rate of Evaporation

Unhardened concrete is in danger of permanent damage when the rate of evaporation exceeds 0.2 lb/sq ft/hr. A weather condition chart can be used to estimate the expected rate of evaporation. The procedure to determine the rate of evaporation is as follows:

1. Find the air temperature and move up to the relative humidity curve.
2. Move right to the concrete temperature.
3. Move down to wind velocity.
4. Move left for rate of evaporation.

For example, if the air temperature is 70°F, the relative humidity is 40%, the concrete temperature is 60°F, and the wind velocity is 20 mph, the rate of evaporation is 0.1 lb/sq ft/hr.

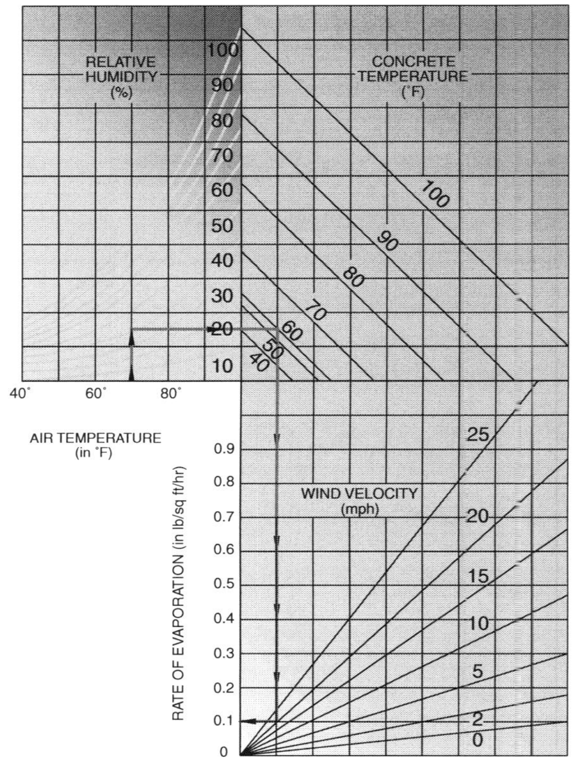

Determine the rate of evaporation when the following conditions exist:

______________________ **1.** Air temperature = 80°F; relative humidity = 40%; concrete temperature = 70°F; wind velocity = 5 mph

______________________ **2.** Air temperature = 55°F; relative humidity = 40%; concrete temperature = 70°F; wind velocity = 5 mph

______________________ **3.** Air temperature = 70°F; relative humidity = 60%; concrete temperature = 80°F; wind velocity = 15 mph

______________________ **4.** Air temperature = 90°F; relative humidity = 80%; concrete temperature = 90°F; wind velocity = 15 mph

______________________ **5.** Air temperature = 60°F; relative humidity = 50%; concrete temperature = 50°F; wind velocity = 25 mph

Projects

1. Air temperature, relative humidity, and wind velocity affect the rate of evaporation from concrete. Average temperature, humidity, and wind velocity records are kept for historical purposes and for use in projecting future weather conditions. Research weather records for your geographic area.

2. Visit a concrete construction job site and identify the curing processes being used. Inquire why these processes were selected for the particular job.

Concrete Principles

Chapter 8

Tool and Equipment Maintenance

Name ______________________________ **Date** ____________________

True-False

T F **1.** Tools should be washed thoroughly by spraying with water and scrubbed with a stiff-bristled brush after daily use.

T F **2.** A flywheel is a protruding mass integrally cast into the crankshaft journal, which partially balances the forces of a reciprocating piston and reduces the load on crankshaft bearing journals.

T F **3.** If concrete comes into contact with a level, the level should be cleaned with a damp cloth.

T F **4.** A piston ring is an expandable split ring used to provide a seal between the piston and cylinder bore.

T F **5.** The intake/compression stroke of a two-stroke cycle engine is the stroke in which the charge is ignited and the burned charge is expelled from the combustion chamber.

T F **6.** Compressed air can be used to remove small grit and dirt from the cooling fins of an engine.

T F **7.** Never wash the foam element of an air cleaner in gasoline or other low flash point solvents.

T F **8.** Dust should be removed from transit levels and laser transit levels using an air compressor.

T F **9.** Engine oil should be drained while the engine is hot.

T F **10.** Cast iron cylinder blocks are lightweight and dissipate heat better than cast aluminum alloy cylinder blocks.

T F **11.** A bearing is used to reduce friction and to maintain clearance between stationary and rotating components of an engine.

T F **12.** Always replace both motor brushes at the same time.

T F **13.** Under normal conditions, air cleaner elements should not be removed from a rammer.

T F **14.** Tape measure blades should be wiped with an oiled cloth to prevent corrosion.

T F **15.** The piston in a two-stroke cycle engine is required to make two complete passes in the cylinder for each operating cycle.

T F **16.** A dual-element air cleaner consists of a foam precleaner and a paper cartridge filter.

T F **17.** A concrete saw is typically powered by a four-stroke cycle engine.

T F **18.** A periodic maintenance schedule is a list of the maintenance tasks and the frequency of when the tasks should be performed.

T F **19.** Overtightened belts do not affect the efficiency or operation of an engine.

T F **20.** The air cleaner of a vibratory plate should be cleaned every 250 hr of operation.

T F **21.** Scraper bars are used to prevent carbon buildup inside the cylinder bore of an engine.

T F **22.** A stiff-bristled brush should be used to remove debris from the cooling fins and cooling air intake openings.

T F **23.** When cleaning a paper cartridge air filter, the choke should be closed or the air intake openings should be covered to prevent debris from entering the engine.

T F **24.** When lubricating the bearings of a vibratory truss screed, as much grease as possible should be added to the shaft pillow bearings.

T F **25.** When cleaning a power buggy, avoid direct water pressure to the engine, hydraulic system, water tank, hoses, labels, and other plastic parts.

T F **26.** Motor-in-head vibrators require considerable on-the-job maintenance.

T F **27.** Undertightened belts can slip and come out of their pulleys causing damage to engines and equipment.

T F **28.** Wash the foam element of an air cleaner in a solution of mild detergent and warm water.

T F **29.** Most rammers use an oil foam element air cleaner.

T F **30.** Ramming system oil is checked through the sightglass located on the back of the ram housing.

T F **31.** The core of a flexible shaft vibrator requires lubrication if it makes a rattling sound when the shaft is shaken.

T F **32.** The bucket of a power buggy should be cleaned at the beginning of each day to prevent concrete buildup.

T F **33.** Petroleum-based oil should not be used to lubricate a vibrator head as the high speed and temperature of the rotating head causes the oil to break down and lose its lubricating ability.

T F **34.** Hydraulic oil can ruin fresh concrete.

T F **35.** Power trowel blades and float pans should be sprayed with a water hose and scrubbed with a stiff-bristled brush.

T F **36.** A troubleshooting chart is a logical listing of problems and recommended actions.

T F **37.** Belt deflection is checked halfway between the clutch pulley and the exciter pulley.

T F **38.** The articulated joint and rear drum of a vibratory roller should be lubricated using a general-purpose grease.

T F **39.** The flexible shaft of a flexible shaft vibrator consists of a braided wire core surrounded by a protective casing.

T F **40.** Leaks in a pneumatic system cannot be located without sophisticated diagnostic equipment.

T F **41.** A gearbox vent allows oil trapped inside the gearbox to escape.

T F **42.** Suspected small leaks in a hydraulic system are located safely by passing a clean piece of paper or cardboard near the suspected leak source.

T F **43.** The cooling fins of a concrete saw should be inspected and cleaned at the end of each day.

T F **44.** The exciter of a pneumatic internal vibrator can stick if the vibrator has not been used for a period of time and internal moisture has slightly rusted the inside of the vibrator.

T F **45.** The outside diameter of an internal vibrator should be monitored closely to prevent a sudden collapse of the head casing or interference with the internal working of the head.

T F **46.** The control arm linkages of a ride-on power trowel are used to steer the machine.

Multiple Choice

______________ **1.** When tightening the tip of an electric motor-in-head vibrator, torque the cap to ___.
- A. 150 ft-lb to 300 ft-lb
- B. 250 ft-lb to 400 ft-lb
- C. 350 ft-lb to 500 ft-lb
- D. 450 ft-lb to 600 ft-lb

______________ **2.** A(n) ___ converts heat energy from combustion of fuel into mechanical energy.
- A. motor
- B. internal combustion engine
- C. carburetor
- D. universal motor

______________ **3.** A ___ converts linear motion of the piston into rotary motion.
- A. cylinder head
- B. cylinder bore
- C. crankshaft
- D. counterweight

_______________ **4.** A ___ is a precision ground surface within which the crankshaft rotates.
A. crankpin journal
B. bearing
C. crankshaft
D. bearing journal

_______________ **5.** A ___ is the part of an armature that connects each winding to insulated copper bars on which brushes ride.
A. field
B. commutator
C. lamination
D. coil

_______________ **6.** The automatic transmission fluid used for exciter lubrication should typically be changed every ___ hr of operation.
A. 100
B. 200
C. 300
D. 500

_______________ **7.** A small engine is an internal combustion engine that is generally rated at ___ HP or less.
A. 3.5
B. 5
C. 15
D. 25

_______________ **8.** The rollover protection structure of a vibratory smooth-drum roller should typically be inspected ___ to ensure that the screws holding the structure in position are tight.
A. daily
B. weekly
C. monthly
D. annually

_______________ **9.** The oil level of the trowel gearbox for a walk-behind power trowel should typically be checked ___.
A. daily
B. weekly
C. monthly
D. annually

_______________ **10.** Air cleaners on concrete saws must be typically inspected and cleaned ___ so that air is allowed to enter the engine.
A. daily
B. weekly
C. monthly
D. annually

_______________ **11.** The ___ is the main structure of a small engine that supports and helps maintain alignment of internal and external components.
A. cylinder block
B. engine block
C. cylinder
D. crankcase

_______________ **12.** ___ current flows in one direction only in a circuit.
A. Alternating
B. Universal
C. Direct
D. all of the above

_______________ **13.** The piston of a four-stroke cycle engine makes ___ complete passes in the cylinder to complete one operating cycle.
A. two
B. three
C. four
D. six

_______________ **14.** Proper belt deflection for a concrete saw is ___.
A. 1⁄16″
B. 1⁄4″
C. 1⁄2″
D. 3⁄4″

_______________ **15.** A flywheel is a(n) ___ disk.
A. zinc
B. cast iron
C. aluminum
D. all of the above

_______________ **16.** A ___ is an engine component that houses and supports the crankshaft.
A. cylinder head
B. cylinder bore
C. crankcase
D. counterweight

_______________ **17.** The ___ is the power trowel component that retains a trowel blade.
A. lifting fork
B. trowel arm
C. cam follower
D. spider

_______________ **18.** One of the basic components of an electric flexible shaft vibrator is the ___.
A. motor
B. shaft
C. head
D. all of the above

_______________ **19.** The ___ is a hole in the engine block that aligns and directs the piston during movement.
- A. cylinder head
- B. cylinder bore
- C. crankcase
- D. counterweight

_______________ **20.** Engine oil for a vibratory smooth-drum roller should typically be replaced after the first ___ hr of operation and every 50 hr thereafter.
- A. 2
- B. 4
- C. 8
- D. 16

_______________ **21.** The ___ consists of the cylinder bore, cooling fins, and valve train components, depending on the engine design.
- A. engine block
- B. crankcase
- C. cylinder block
- D. cylinder head

_______________ **22.** The ___ is a cast aluminum or cast iron engine component fastened to the end of a cylinder block farthest from the crankshaft.
- A. cylinder head
- B. cylinder bore
- C. crankcase
- D. counterweight

_______________ **23.** The maximum belt deflection for a ride-on power trowel belt should be ___ at the center of the span between the pulleys.
- A. 1⁄16″ to 1⁄8″
- B. 1⁄8″ to 1⁄4″
- C. 1⁄2″ to 3⁄4″
- D. 3⁄4″ to 1″

_______________ **24.** The ___ stroke of a two-stroke cycle engine is the stroke in which the charge is ignited and the burned charge is expelled from the combustion chamber.
- A. power
- B. exhaust
- C. power/exhaust
- D. none of the above

_______________ **25.** The fluid level of a hydraulic external vibrator should typically be inspected ___ to ensure that an adequate supply is present in the hydraulic system.
- A. daily
- B. weekly
- C. monthly
- D. annually

______________ **26.** The condition of the air cleaner indicator of a sheepsfoot roller should typically be checked ___ to ensure proper operation of the roller.
A. daily
B. weekly
C. monthly
D. annually

______________ **27.** A ___ is a disk that is mounted at one end of the crankshaft to provide inertia for the engine.
A. flywheel
B. counterweight
C. crankshaft
D. bearing

______________ **28.** The ___ stroke of a four-stroke cycle engine is the stroke in which heat energy produced by the explosive charge is converted to mechanical energy.
A. intake
B. compression
C. power
D. exhaust

______________ **29.** The ___ stroke of a four-stroke cycle engine is the stroke in which the burned charge is expelled from the cylinder.
A. intake
B. compression
C. power
D. exhaust

______________ **30.** A ___ provides the required air-fuel mixture to the combustion chamber based on engine operating speed and load.
A. crankcase
B. carburetor
C. crankcase breather
D. cylinder bore

______________ **31.** Ramming system oil must typically be changed after the first 50 hr of rammer operation and every ___ hr thereafter.
A. 50
B. 100
C. 300
D. 500

______________ **32.** The hydraulic system, including all lines and connections, should typically be inspected each ___.
A. day
B. week
C. month
D. year

_______________ **33.** Ramming shoe bolts should typically be checked and tightened after the first 5 hr of operation and every ___ thereafter.
A. day
B. week
C. month
D. year

_______________ **34.** A ___ is an extension of the crankshaft that allows an engine to transmit power to an application such as a concrete saw.
A. crankpin
B. power take-off
C. counterweight
D. piston

_______________ **35.** After installing a new belt on a concrete saw, the belt tension should typically be checked and adjusted after ___ hr of use.
A. 1
B. 4
C. 8
D. none of the above

_______________ **36.** The ___ stroke of a four-stroke cycle engine is the stroke in which the air-fuel mixture is introduced into the cylinder.
A. intake
B. compression
C. power
D. exhaust

_______________ **37.** A(n) ___ is the part of a motor that produces a rotating magnetic field.
A. armature
B. brush
C. commutator
D. field

Completion

_______________ **1.** A(n) ___ is a machine that converts a form of energy into mechanical force.

_______________ **2.** A(n) ___ is a gear located on the crankshaft that is used to drive other parts of an engine.

_______________ **3.** The ___ is the part of an internal combustion engine that includes components required to control the flow of gases into and out of the combustion chamber.

_______________ **4.** Approximately ___% of the energy released from gasoline is converted into work.

_______________ **5.** ___ is a force that causes rotation in a motor.

_______________ **6.** New brushes should be installed if their length is less than ___″.

_______________ **7.** The belt of a vibratory plate should typically be checked after the first ___ hr of operation and every 50 hr thereafter.

______________ **8.** If a vibratory roller drifts in either direction when the forward/reverse control is in neutral, the ___ cable must be adjusted.

______________ **9.** A(n) ___ is a machine that converts electrical energy into rotating mechanical force.

______________ **10.** The engine component in which the cylinder block and cylinder head are cast as a single unit is a(n) ___.

______________ **11.** A(n) ___ is the part of a universal motor that provides contact between the external power source and the commutator.

______________ **12.** The hydraulic fluid of a power buggy should typically be checked every ___ and fluid should be added if the level is less than one-half full in the sightglass.

______________ **13.** Hydraulic fluid of a vibratory roller is checked ___ by viewing the level through the sightglass.

______________ **14.** The shaft core of an electric flexible shaft vibrator should typically be lubricated after every ___ hr of use.

______________ **15.** A trowel ___ is a sealed container that has an input shaft and an output shaft and uses a set of mating gears.

______________ **16.** A cooling ___ is an integral thin, cast strip designed to provide efficient air circulation and dissipation of heat away from the engine cylinder and into the surrounding air.

______________ **17.** A(n) ___ motor is a motor that can be operated on either direct current (DC) or alternating current (AC).

______________ **18.** A(n) ___ control knob is a power trowel component that adjusts the angle of the trowel blades.

______________ **19.** ___ is the property of matter by which it remains in uniform motion unless acted upon by some external force.

______________ **20.** The crankcase ___ relieves crankcase pressure created by the reciprocating motion of the piston during engine operation.

______________ **21.** A(n) ___ is filler material between the cylinder block and cylinder head used to seal the combustion chamber.

______________ **22.** A(n) ___ journal is a precision ground surface that provides a rotating pivot point to attach the connecting rod and crankshaft.

______________ **23.** ___ current reverses its direction of flow in a circuit twice per cycle.

______________ **24.** The no-load engine speed of a vibratory truss screed should be ___ rpm ±100 rpm.

______________ **25.** The engine component that slides back and forth in the cylinder bore by forces produced during the power stroke is a(n) ___.

______________ **26.** A(n) ___ transfers motion from the piston to the crankshaft and functions as a lever arm.

______________ **27.** The ___ stroke is the stroke of a four-stroke cycle engine in which the air-fuel mixture is condensed within the cylinder.

______________ **28.** ___ is the process of removing air that may have entered the hydraulic system.

______________ **29.** ___ is the systematic elimination of various parts of a system or process to locate a defective or malfunctioning component.

______________ **30.** The fuel tank of a concrete saw should typically be drained and cleaned per ___ recommendations.

______________ **31.** A(n) ___ is the volume of compressed air-fuel mixture trapped inside the combustion chamber ready for ignition.

______________ **32.** ___ maintenance is the tasks completed at specific intervals to prevent breakdowns and production inefficiency.

______________ **33.** A(n) ___ transfers power between two pulleys.

______________ **34.** The vibrator head of an electric flexible shaft vibrator should be lubricated using a good ___-based synthetic oil.

______________ **35.** A(n) ___ is the weighted part of a pneumatic internal vibrator that rotates rapidly to produce vibrations.

______________ **36.** Ride-on power trowel belts should typically be inspected every ___ hr or more frequently in adverse conditions.

______________ **37.** A(n) ___ is a component of a pneumatic system that supplies a mist of special oil into a compressed air line to lubricate pneumatic tools and internal motor parts.

______________ **38.** A(n) ___ is the failure of a system, equipment, or component to operate as designed.

______________ **39.** A(n) ___ system transmits energy using a gas such as compressed air.

______________ **40.** A(n) ___ system transmits energy using a fluid under pressure.

______________ **41.** The part of the motor that rotates the motor shaft and delivers the work is the ___.

Short Answer

1. List the recommended procedure for removing dents or large nicks in the edges of spreaders, rakes, or shovel blades.

____

____

____

2. List three concrete hand tools that are commonly used and cite two maintenance requirements for each.

____

____

____

3. Explain why regular periodic maintenance is necessary for efficient operation of equipment.

4. Describe the operation of a four-stroke cycle engine.

5. Summarize the procedure for cleaning a dual-element air cleaner.

6. Describe how a belt is inspected for proper tension.

7. Describe the operation of a two-stroke cycle engine.

8. Describe the procedure for draining and replacing engine oil.

____

____

____

____

9. List four maintenance tasks that should be performed on a daily basis for a vibratory truss screed.

____

____

____

____

10. Identify two maintenance tasks that should be performed after each use of a walk-behind power trowel.

____

____

____

____

Matching

Pistons

______	**1.** Piston head	**A.** Portion of a piston closest to the crankshaft that helps align the piston as it moves within the cylinder bore
______	**2.** Piston pin	**B.** Top surface of the piston, which is subject to extreme forces and heat during engine operation
______	**3.** Skirt	**C.** Recessed area located around the perimeter of a piston that is used to retain a piston ring
______	**4.** Ring groove	**D.** Hollow shaft that connects the small end of a connecting rod to the piston

Piston Rings

______	**1.** Compression rings	**A.** Further seals the combustion chamber and wipes cylinder walls clean of excess oil
______	**2.** Wiper rings	**B.** Seals the combustion chamber from any leakage during combustion
______	**3.** Oil rings	**C.** Wipes excess oil from the cylinder wall during piston movement

Walk-Behind Power Trowels

______ **1.** Trowel blade

______ **2.** Oil sight glass

______ **3.** Fill/drain plug

______ **4.** Lifting fork

______ **5.** Pivot point

______ **6.** Trowel gearbox

______ **7.** Trowel arm

______ **8.** Spider

______ **9.** Gearbox vent

______ **10.** Trowel arm grease fittings

______ **11.** Cam follower grease fittings

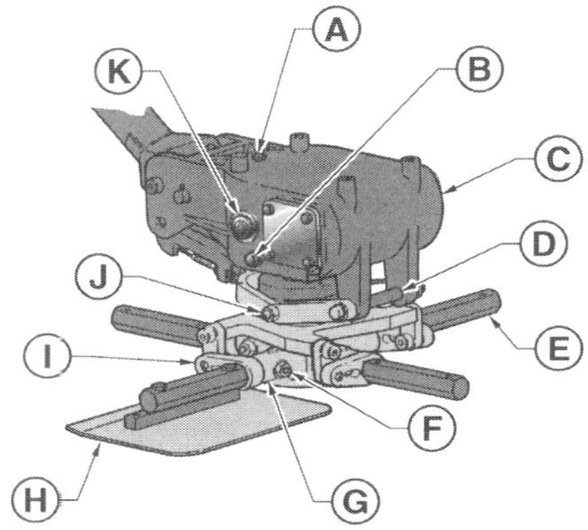

Cylinder Blocks

______ **1.** Head gasket

______ **2.** Cylinder bore

______ **3.** Spark plug

______ **4.** Cooling fin

______ **5.** Cylinder block

______ **6.** Cylinder head

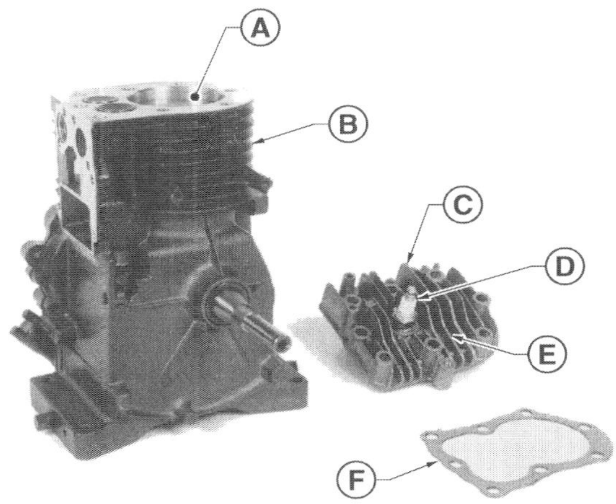

Crankshafts

_______________ 1. Crankcase

_______________ 2. Crankshaft

_______________ 3. Counterweight

_______________ 4. Crankpin journal

_______________ 5. Power take-off

_______________ 6. Connecting rod

_______________ 7. Bearing

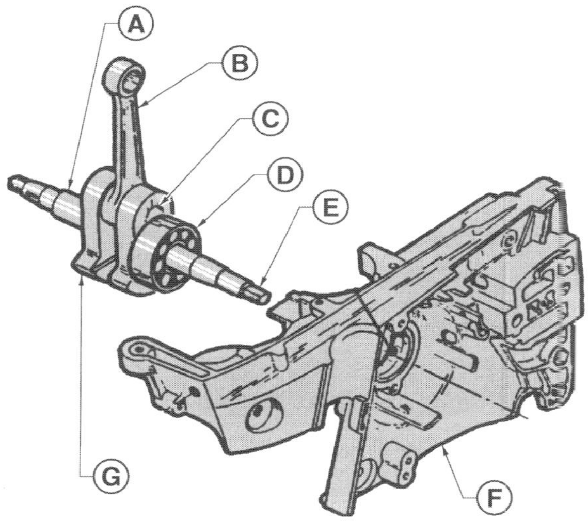

Piston Components

_______________ 1. Piston head

_______________ 2. Compression ring

_______________ 3. Skirt

_______________ 4. Wiper ring

_______________ 5. Ring groove

_______________ 6. Oil ring

_______________ 7. Piston pin bore

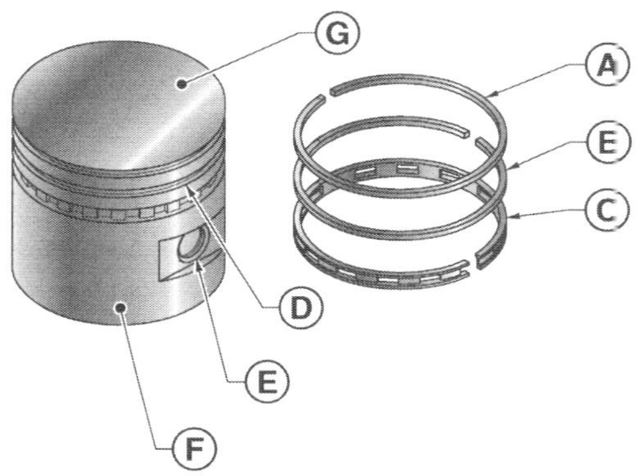

Connecting Rods

_______________ 1. Piston pin

_______________ 2. Connecting rod

_______________ 3. Rod cap screw

_______________ 4. Crankpin journal bearing surface

_______________ 5. Piston

_______________ 6. Rod cap

_______________ 7. Spring clip

_______________ 8. Piston pin bearing surface

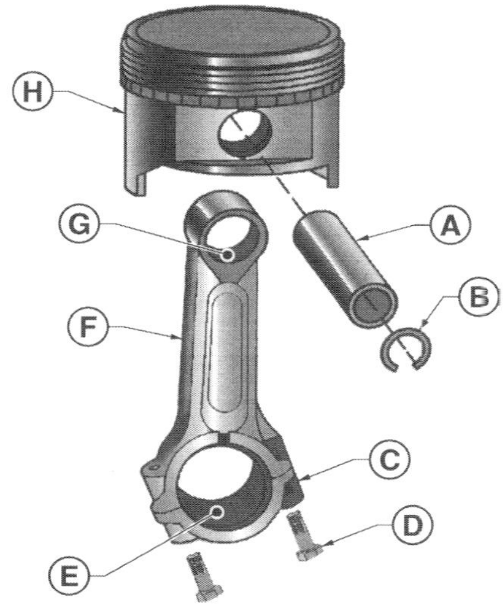

Universal Motors

______ **1.** Commutator

______ **2.** O-ring

______ **3.** Field

______ **4.** Brush

______ **5.** Spring

______ **6.** Brush holder

______ **7.** Armature

______ **8.** Brush cap

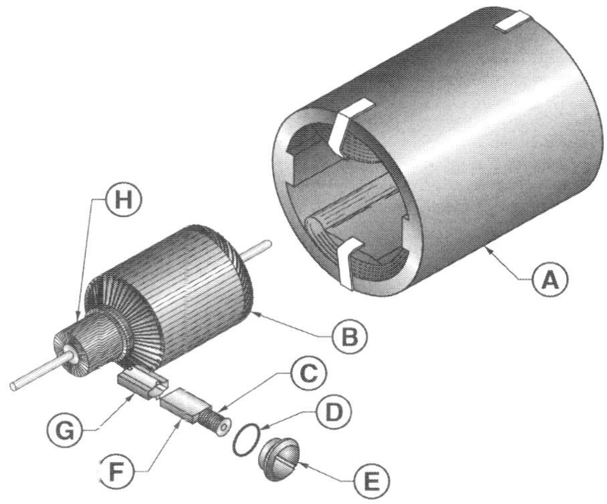

Engine Oil Maintenance

______ **1.** Crankcase

______ **2.** Drain plug

______ **3.** Oil fill plug

______ **4.** Dipstick

______ **5.** O-ring

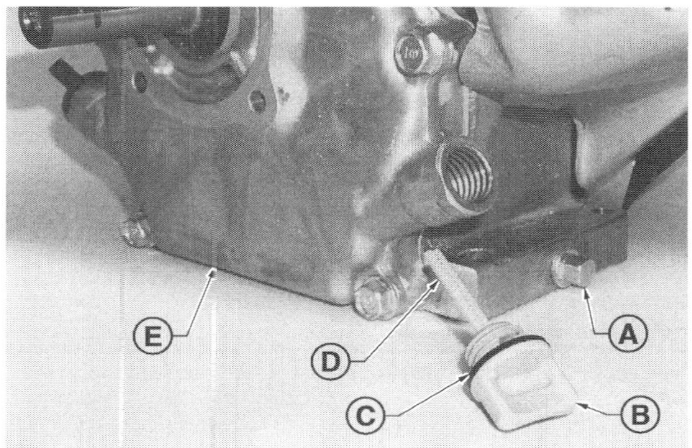

Concrete Saw Fuel Systems

______ **1.** Fuel tank

______ **2.** Fuel tank opening

______ **3.** Fuel line

______ **4.** Rubber jacket

______ **5.** Vent valve

______ **6.** Fuel passage

______ **7.** Fuel cap

______ **8.** O-ring

______ **9.** Vent Hole

______ **10.** Weighted fuel filter

______ **11.** Flexible fuel suction line

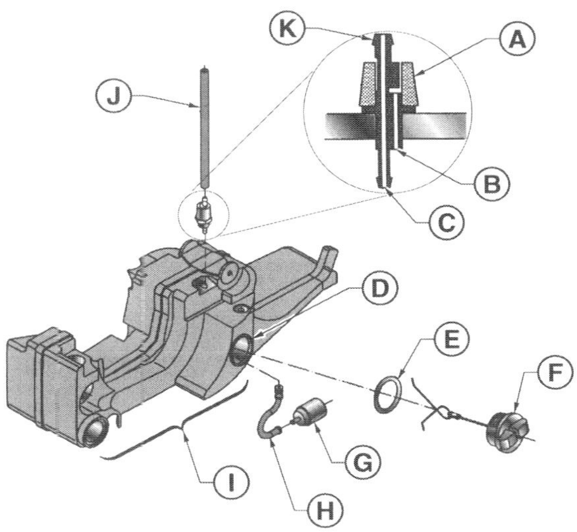

Activities

Engine Displacement

Displacement is the volume that a piston displaces in an engine when it travels from the top of the cylinder bore to the bottom of the bore. When the cylinder bore diameter and length of stroke are known, the displacement is calculated by applying the formula:

$D = 0.7854 \times B^2 \times S$

where

D = displacement (in cu in.)
0.7854 = constant
B = cylinder bore diameter (in in.)
S = stroke (in in.)

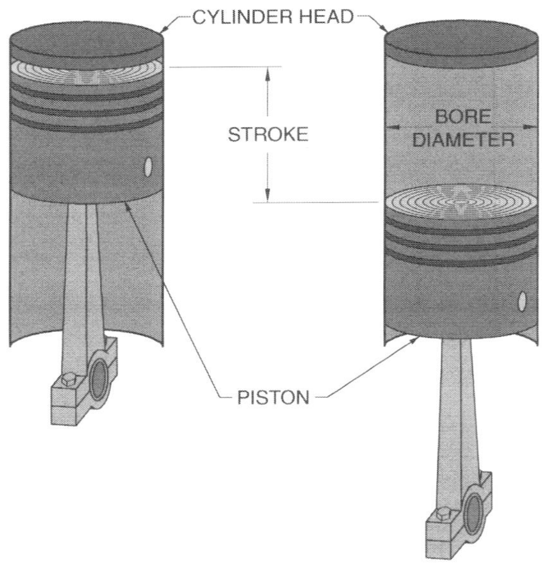

For example, what is the displacement of a single-cylinder engine that has a 2.5″ cylinder bore diameter and a 1.75″ stroke?

$D = 0.7854 \times B^2 \times S$
$D = 0.7854 \times (2.5 \times 2.5) \times 1.75$
$D = 0.7854 \times 6.25 \times 1.75$
$D = 8.59$ cu in.

Determine the displacement of the following engines:

______________ **1.** Cylinder bore diameter = 2.625″; stroke = 1.45″

______________ **2.** Cylinder bore diameter = 2.45″; stroke = 1.875″

______________ **3.** Cylinder bore diameter = 2.75″; stroke = 1.95″

______________ **4.** Cylinder bore diameter = 3.125″; stroke = 3.25″

______________ **5.** Cylinder bore diameter = 3.25″; stroke = 3.75″

Projects

1. Inspect a piece of power equipment commonly used on a job site. Note the engine rating and whether it is a two-stroke cycle or four-stroke cycle engine. Check the oil level and inspect the air cleaner. Identify how the engine is started and stopped and locate the controls for the engine.
2. Obtain a copy of an operator's manual for equipment on your job site. Review the manual for information regarding maintenance, including the periodic maintenance charts and troubleshooting charts.

Concrete Principles

Chapter 9

Concrete Quality, Testing, and Repair

Name ______________________________ **Date** ____________________

True-False

T F **1.** Mix water containing sulfate concentrations of 1% or less can be used safely in concrete mixtures.

T F **2.** Cement fineness impacts the strength-gaining properties of concrete, especially early strength development.

T F **3.** When obtaining aggregate samples, select samples as the aggregate is being transported around the supplier's site to ensure a random selection.

T F **4.** Aggregate must be clean and free from any foreign material, especially organic matter.

T F **5.** Admixtures should be added to portland cement prior to adding mix water.

T F **6.** Crushed stone has a texture similar to gravel.

T F **7.** Accelerators are especially useful in cold weather conditions to allow concrete to set before it freezes.

T F **8.** A concrete mixture containing many rough, flat, or elongated particles requires additional sand for cohesiveness.

T F **9.** Concrete strength will increase when using water-reducing admixtures, provided that the water content in the mixture is reduced and cement content and slump remain constant.

T F **10.** Superplasticizers remove some of the entrained air in a mixture.

T F **11.** Wood floats should be used for air-entrained concrete since they do not tear the surface.

T F **12.** A 10 lb sample of fine aggregate or a 50 lb sample of coarse aggregate should be obtained for each 50 ton field sample size.

T F **13.** Patching mixes have a long pot life.

T F **14.** To ensure high-quality concrete, it should be placed with the lowest slump possible that can be handled and consolidated by equipment and workers on the job.

T F **15.** Gasoline-powered or electric portable mixers are used on mid-size jobs such as a section of sidewalk or concrete steps.

T F **16.** The fineness modulus of aggregate that is obtained from different sources should not vary more than 0.2.

T F **17.** Liquid admixtures are added to mix water or sprayed on nonabsorptive or saturated aggregate.

T F **18.** Most set-retarding admixtures increase the amount of water needed in a concrete mixture.

T F **19.** Calcium chloride reacts with aluminum and creates heat and gas bubbles.

T F **20.** When using coloring admixtures, the color of the portland cement does not have an effect on the brightness or vividness of the concrete.

T F **21.** Air detrainers should be used in cold environments.

T F **22.** Performance specifications place the responsibility of proper mixture design on the concrete supplier.

T F **23.** Concrete for small projects, such as patching or sidewalk repair, is best batched and mixed in a wheelbarrow.

T F **24.** Powdered admixtures should be added to portland cement before introducing them into the mixture.

T F **25.** Set-retarding admixtures are useful in hot weather conditions when concrete sets so rapidly that it cannot be finished properly.

T F **26.** Calcium chloride is added directly to a concrete mixture.

T F **27.** Coloring admixtures are thoroughly blended with dry portland cement or a dry concrete mixture prior to adding water to ensure a uniform color of cement paste throughout the mixture.

T F **28.** When using a set-retarding admixture, the strength of concrete increases during the first three days after placement.

T F **29.** Rock pockets and honeycomb are typically found along edges or at corners of concrete when stripping forms.

T F **30.** Never add calcium chloride to concrete in its dry form as it might not dissolve properly.

T F **31.** Cracking can be reduced by decreasing the amount of mix water used and producing lower slump concrete.

T F **32.** Calcium chloride does not have an effect on steel reinforcement.

T F **33.** When calcium chloride is added to a concrete mixture in proper proportions, it does not decrease the freezing point of concrete to any great extent.

T F **34.** Concrete containing pozzolan usually requires additional mix water to achieve the same slump as concrete that does not contain pozzolan.

T F **35.** Shrinkage cracks usually form after concrete is completely hardened.

T F **36.** A coloring admixture imparts a desired color to concrete.

T F **37.** Slabs should not be strongly restrained at the perimeter and should not be bonded to footing pads or other structures beneath them.

T F **38.** Non-air-entrained concrete bleeds more water than air-entrained concrete.

T F **39.** Blistering is most likely to occur with non-air-entrained concrete.

T F **40.** Dusting may be caused by the use of unvented heaters in enclosed areas during cold weather construction.

T F **41.** Spalling is a deeper penetration of the slab than scaling or blistering.

T F **42.** Steel trowels produce an excessively dark surface if not properly used when dry packing concrete.

T F **43.** Shrinkage cracks that occur in unhardened concrete may be closed using a tamper.

T F **44.** Very severe scaling involves loss of coarse aggregate and mortar, generally to a depth greater than ¾″.

T F **45.** Grout should not be used to repair cracks and joints until concrete has cooled completely.

T F **46.** Wet or high-slump concrete mixtures have a slump ranging from 2″ to 4″.

T F **47.** When epoxy resins are used to repair concrete, additional heat must be applied for curing.

T F **48.** Light scaling does not expose coarse aggregate.

T F **49.** When applying cures to slabs, the surface, edges, and joints must be properly cured.

T F **50.** When measuring the concrete slump, an accurate measurement is obtained by measuring to the center of the concrete.

T F **51.** Spalls are typically found in the middle of concrete slabs where bleedwater is slow to evaporate.

T F **52.** Dusting is commonly attributed to excessive bleedwater dissipation and rapid drying of the surface due to poor curing conditions.

T F **53.** A bonding agent is applied to a surface using a hand sprayer.

T F **54.** When making repairs, it is better to remove too much concrete than too little because the affected area continues to disintegrate if it is not removed.

T F **55.** When performing a unit weight test, the scale should be accurate to within 0.1% of the test load.

T F **56.** The gravimetric method for conducting an air content test is a practical method for field testing concrete.

T F **57.** Project specifications commonly require six test cylinders to be made.

T F **58.** A beam test specimen must be thoroughly dry before performing a beam test.

T F **59.** The pressure method for conducting an air content test is a practical method of testing all concrete except concrete made with highly porous and lightweight aggregate.

T F **60.** Field test concrete samples for a compression test should be taken at three or more regular intervals during the discharge of the batch.

T F **61.** A higher modulus of rupture indicates a lower load-carrying capacity of concrete.

T F **62.** Compression tests performed early determine whether forms may be stripped or whether the structure can be put under load.

T F **63.** Compression test specimens should not be moved for at least 24 hr.

T F **64.** When molding samples for a compression test, either vibration or rodding can be used to consolidate the samples.

T F **65.** The length of a core extraction specimen should be as close as possible to twice the diameter of the core.

T F **66.** A beam test specimen should remain in the mold between 24 hr ±8 hr.

T F **67.** When molding beam test specimens, rodding can be used for concrete with a slump between 1″ and 3″.

T F **68.** Under favorable curing conditions, the full strength of concrete using pozzolan as a replacement for some of the portland cement is generally higher than the strength of concrete using portland cement only.

T F **69.** Natural pigments and earth colors are durable colors for coloring admixtures.

Multiple Choice

______________ **1.** ___ is the level of clarity or purity in a liquid.
A. Turbidity
B. Pureness
C. Mudity
D. none of the above

______________ **2.** The fineness modulus for fine aggregate is between ___.
A. 1.1 and 1.7
B. 1.8 and 2.3
C. 2.3 and 3.1
D. 3.1 and 3.7

______________ **3.** Tricalcium silicate and dicalcium silicate provide ___ characteristics of concrete.
A. hydration heat
B. sulfate resistance
C. strength-developing
D. all of the above

______________ **4.** ___ are substances other than water, aggregate, or portland cement that are added to concrete to modify its properties.
A. Clinkers
B. Admixtures
C. Fillers
D. all of the above

__________ **5.** Type ___ cement, which is used in high-early-strength concrete, is typically ground much finer than other types of cement.
A. I
B. II
C. III
D. IV

__________ **6.** An accelerator such as ___ produces rapid setting and considerable concrete strength within a few hours of placement.
A. sodium silicate
B. high alumina cement
C. triethanolamine
D. all of the above

__________ **7.** Properties such as ___ can be impacted using an admixture or combination of admixtures.
A. workability
B. freeze/thaw resistance
C. retardation of setting time
D. all of the above

__________ **8.** A(n) ___ admixture is added to a concrete mixture to improve the impermeability of hardened concrete.
A. bonding
B. air-entraining
C. dampproofing
D. water-reducing

__________ **9.** ___ contributes to high-early-strength and high heat development in cement, but can result in poor sulfate resistance and significant volume change.
A. Tricalcium silicate
B. Dicalcium silicate
C. Tricalcium aluminate
D. Tetracalcium aluminoferrite

__________ **10.** The amount of coloring admixture in a mixture ranges from ___ by weight of the cement used in the mixture.
A. 1% to 2%
B. 2% to 5%
C. 2% to 10%
D. 10% to 15%

__________ **11.** The ___ process is a process in which concrete ingredients are completely mixed before unloading them into a truck to transport concrete to the job site.
A. transit-mixed
B. central-mixed
C. shrink-mixed
D. none of the above

_______________ **12.** The ___ is a factor that is obtained by adding the total percentages by weight of an aggregate sample retained on sieves No. 4, 8, 16, 30, 50, and 100 and then dividing the total by 100.

A. fineness modulus
B. turbidity factor
C. sieve factor
D. grading level

_______________ **13.** ___ is used as a filler in cement and should be kept to a minimum.

A. Tricalcium silicate
B. Dicalcium silicate
C. Tricalcium aluminate
D. Tetracalcium aluminoferrite

_______________ **14.** A(n) ___ admixture is added to a concrete mixture to reduce setting time and improve the early strength of concrete.

A. accelerating
B. set-retarding
C. water-reducing
D. air-entraining

_______________ **15.** The amount of set-retarding admixture used in a concrete mixture should be less than ___% by weight of the cement.

A. 0.03
B. 0.3
C. 3
D. 30

_______________ **16.** The slump increase resulting from the addition of a superplasticizer lasts approximately ___.

A. ½ hr to 1 hr
B. 1 hr to 3 hr
C. 3 hr to 5 hr
D. 5 hr to 8 hr

_______________ **17.** Cement containing a large percentage of ___ has a high heat of hydration and results in concrete that has high early-strength.

A. tricalcium silicate
B. dicalcium silicate
C. tricalcium aluminate
D. tetracalcium aluminoferrite

_______________ **18.** A(n) ___ admixture is added to a concrete mixture to reduce the amount of water needed to produce the desired mixture.

A. accelerating
B. set-retarding
C. water-reducing
D. air-entraining

_______________ **19.** The most common type of coarse aggregate is ___.
A. gravel
B. blast furnace slag
C. crushed stone
D. all of the above

_______________ **20.** Concrete exposed to repeated freeze/thaw cycles should contain ___% ± ½% entrained air.
A. 2
B. 4
C. 6
D. 8

_______________ **21.** A sulfate concentration of 1% reduces concrete strength by ___%.
A. 1
B. 2
C. 5
D. 10

_______________ **22.** Aggregate comprises ___ of the total volume of concrete.
A. 16% to 18%
B. 26% to 38%
C. 48% to 56%
D. 66% to 78%

_______________ **23.** A common type of pozzolan is ___.
A. fly ash
B. pulverized brick
C. volcanic glass
D. all of the above

_______________ **24.** A(n) ___ admixture is added to concrete to extend its setting time.
A. accelerating
B. set-retarding
C. water-reducing
D. air-entraining

_______________ **25.** Calcium chloride should not be used with Type ___ cement.
A. II
B. III
C. IV
D. V

_______________ **26.** A(n) ___ coloring admixture hardens to a clear blue-gray color.
A. phthalocyanine blue
B. carbon black
C. ultraviolet orange
D. phthalocyanine green

_______________ **27.** Plasticizers typically reduce the amount of water needed in a mixture by ___ with only a slight increase in slump.
A. 5% to 10%
B. 10% to 15%
C. 15% to 20%
D. 20% to 25%

_______________ **28.** Air bubbles that are intentionally incorporated into concrete range in size from ___ diameter.
A. 10 μm to 1000 μm
B. 100 μm to 5000 μm
C. 1000 μm to 5000 μm
D. 5000 μm to 10,000 μm

_______________ **29.** A(n) ___ admixture allows less mix water to be used to produce concrete of a desired slump while retarding the set of the concrete.
A. accelerating
B. set-retarding
C. water-reducing
D. water-reducing, set-retarding

_______________ **30.** Water-reducing admixtures are used in small amounts, generally less than ___% of the weight of the cement.
A. 0.03
B. 0.3
C. 3
D. 30

_______________ **31.** An internal vibrator should be inserted into the concrete at approximately ___ intervals, depending on vibration speed and slump of the concrete.
A. 2″ to 4″
B. 4″ to 6″
C. 6″ to 8″
D. 8″ to 10″

_______________ **32.** A(n) ___ admixture is a foaming substance used to add microscopic air bubbles to concrete.
A. accelerating
B. set-retarding
C. water-reducing
D. air-entraining

_______________ **33.** Unless otherwise specified, concrete should have a ___″ slump or less if vibration is used to consolidate fresh concrete.
A. 1
B. 2
C. 3
D. 4

______________ **34.** The ___ effect is a condition of concrete in which concrete hardens at the surface but remains soft in the middle.
A. turbidity
B. hexatropic
C. setting
D. none of the above

______________ **35.** A common bonding admixture is ___.
A. acrylic
B. latex
C. polyvinyl
D. all of the above

______________ **36.** ___ specifications are usually written by a specification writer.
A. Project
B. Performance
C. Prescriptive
D. none of the above

______________ **37.** Commercial ready-mixed concrete plants produce concrete that is ___.
A. transmit-mixed
B. shrink-mixed
C. central-mixed
D. all of the above

______________ **38.** Most specifications allow for ± ___% in weight of cement or weight or volume of water.
A. 1
B. 2
C. 3
D. 4

______________ **39.** Only ___% of cement particles in a normal concrete mixture are hydrated.
A. 16
B. 26
C. 46
D. 66

______________ **40.** Concrete consistency is measured by conducting a ___ test.
A. beam
B. slump
C. nuclear
D. compression

______________ **41.** Air content in a concrete mixture decreases when using ___.
A. more cement
B. higher temperatures
C. a pozzolan admixture
D. all of the above

_______________ **42.** A normal concrete mixture of cement, aggregate, and water contains approximately ___ entrapped air.
- A. 1% to 2%
- B. 3% to 4%
- C. 5% to 6%
- D. 7% to 8%

_______________ **43.** ___ is the distortion of a flat concrete surface or member into a curved shape due to creep or differences in temperature or moisture content in areas adjacent to its opposite faces.
- A. Scaling
- B. Spalling
- C. Curling
- D. Bending

_______________ **44.** ___ air is the system of microscopic bubbles intentionally incorporated into a concrete mixture.
- A. Entrapped
- B. Entrained
- C. Entranced
- D. all of the above

_______________ **45.** Maximum size aggregate should not exceed ___ of the minimum clear space between steel reinforcement.
- A. one-eighth
- B. one-fourth
- C. three-eighths
- D. three-fourths

_______________ **46.** In general, air content in a concrete mixture increases with an increase in ___.
- A. slump
- B. fine aggregate content
- C. water-cement ratio
- D. all of the above

_______________ **47.** ___ air is not intentionally incorporated into the concrete mixture and leaves voids in concrete if not properly removed.
- A. Entrapped
- B. Entrained
- C. Entranced
- D. all of the above

_______________ **48.** Maximum size aggregate should not exceed ___ of the narrowest dimension between the sides of forms.
- A. one-fifth
- B. one-fourth
- C. one-third
- D. one-half

________________ **49.** Large spalls may be circular or ___ in shape and are more than ¾″ deep and over 5⅝″ in diameter.

A. square
B. oval
C. rectangular
D. pentagonal

________________ **50.** ___ is the irregular raising of a thin layer of cement paste at the surface of a slab, which appears during or soon after completion of finishing.

A. Scaling
B. Spalling
C. Blistering
D. Dusting

________________ **51.** Excessive bleedwater is a condition where a large quantity of water continues to migrate to the ___ after concrete is placed.

A. edges
B. bottom
C. surface
D. all of the above

________________ **52.** Severe scaling involves loss of mortar 3⁄16″ to ⅜″ deep with some loss of mortar around coarse aggregate extending ___ deep.

A. ⅛″ to ½″
B. ⅜″ to ¾″
C. ⅜″ to 2″
D. 2″ to 6″

________________ **53.** Small spalls are typically a circular depression not greater than ___″ deep or 5⅝″ in any direction.

A. ¼
B. ½
C. ¾
D. 1

________________ **54.** ___ is used to repair bug holes, pitting, and light honeycomb and results in an even texture across the surface.

A. Dry packing
B. Sack rubbing
C. Epoxy injection
D. Grinding

________________ **55.** A temperature of ___ should be maintained for the 24 hr period following the molding process.

A. 20°F to 40°F
B. 40°F to 60°F
C. 60°F to 80°F
D. none of the above

_______________ **56.** Popouts range in diameter from ___.
- A. ⅛″ to ½″
- B. ⅜″ to 2″
- C. ½″ to ¾″
- D. 2″ to 6″

_______________ **57.** Pozzolan replaces ___ of the cement by weight.
- A. 5% to 10%
- B. 5% to 30%
- C. 15% to 35%
- D. 25% to 50%

_______________ **58.** Concrete samples for strength tests must be a minimum of ___.
- A. 1 cu ft
- B. 1 cu yd
- C. 2 cu ft
- D. 2 cu yd

_______________ **59.** Medium scaling exposes coarse aggregate and involves the loss of mortar ___ deep.
- A. 1⁄16″ to ⅛″
- B. 3⁄16″ to ⅜″
- C. ⅜″ to ¾″
- D. ¾″ to 1½″

_______________ **60.** Standard test cylinder molds or compression tests are made from ___.
- A. sheet metal
- B. plastic
- C. nonabsorptive coated cardboard
- D. all of the above

_______________ **61.** When conducting a core extraction test, the core diameter must be at least ___ times the maximum size of aggregate in the mixture.
- A. 2
- B. 3
- C. 4
- D. 6

_______________ **62.** Most specifications allow for ± ___% in aggregate weight.
- A. 1
- B. 2
- C. 3
- D. 4

_______________ **63.** Steel reinforcement should be placed in the top ___ of the slab to counteract curling action.
- A. one-fourth
- B. one-third
- C. one-half
- D. none of the above

________________ **64.** ___ is the development of a light powdery material on the surface of hardened concrete.
A. Scaling
B. Spalling
C. Blistering
D. Dusting

________________ **65.** ___ is the process of using carborundum stones to remove hardened debris from a concrete surface.
A. Dry packing
B. Sack rubbing
C. Epoxy injection
D. Grinding

________________ **66.** When removing concrete prior to dry packing, concrete should be removed to a minimum depth of ___″, regardless of the depth of the void.
A. ⅛
B. ½
C. 1
D. 4

________________ **67.** The results of a rebound hammer test should only be used for comparative purposes since rebound of the plunger is affected by the ___.
A. moisture content of the concrete
B. surface finish
C. aggregate type
D. all of the above

________________ **68.** When filling a slump cone, each layer must be rodded ___ times using a tamping rod.
A. 15
B. 20
C. 25
D. 30

________________ **69.** ___ indicates the ability of concrete to resist failure under loads that cause a concrete member to bend or flex.
A. Turbidity
B. Compressive strength
C. Flexural strength
D. none of the above

________________ **70.** Low- or medium-slump mixtures have a slump ranging from ___.
A. 1″ to 2″
B. 2″ to 4″
C. 4″ to 6″
D. 6″ to 8″

______________ **71.** Light-colored areas in a concrete slab are caused by ___.
A. excessive bleedwater
B. troweling too early
C. plastic sheeting used for curing touching the concrete
D. all of the above

______________ **72.** Bonding agents are usually organic water emulsions including ___.
A. polyvinyl chloride
B. rubber
C. acrylics
D. all of the above

______________ **73.** The removal of a slump cone should take ___ sec ± 2 sec.
A. 5
B. 15
C. 30
D. 45

______________ **74.** An air content test standardized by the ASTM is the ___ method.
A. gravimetric
B. volumetric
C. pressure
D. all of the above

______________ **75.** When performing a beam test, a load is applied to the beam at a rate of ___ lb per min.
A. 600
B. 1200
C. 1800
D. 2400

______________ **76.** Stiff or low-slump mixtures typically have a slump of ___″ maximum.
A. 1
B. 2
C. 3
D. 4

______________ **77.** ___ is a mixture of portland cement and water, with or without fine aggregate, which is mixed to a pourable consistency without separation of ingredients.
A. Epoxy
B. Grout
C. Mortar
D. none of the above

______________ **78.** When preparing a beam test specimen, vibration must be used for concrete with a slump less than ___″.
A. 1
B. 3
C. 4
D. 5

_______________ **79.** Bonding agents are generally added in proportions equivalent to ___ by mass of the cement.
A. 1% to 2%
B. 3% to 10%
C. 5% to 20%
D. 15% to 30%

_______________ **80.** Dark-colored areas in concrete are caused by ___.
A. non-uniform mixing of calcium chloride in a mixture
B. excessive power troweling after adding calcium chloride to a mixture
C. higher cement content or lower water content when compared to an adjacent area
D. all of the above

_______________ **81.** A slump test should be completed in approximately ___ min from the collection of the first sample until the measurement is taken.
A. 7
B. 11
C. 22
D. 34

_______________ **82.** The equipment needed to perform the unit weight test includes a sturdy container of known volume and a(n) ___.
A. accurate scale or balance
B. tamping rod
C. strikeoff plate
D. all of the above

_______________ **83.** For concrete containing aggregate up to ___″ nominal size, a $6 \times 6 \times 20$ beam mold should be used.
A. 2
B. 3
C. 5
D. 6

_______________ **84.** When molding compression test specimens, concrete with a slump greater than ___″ must be rodded.
A. 2
B. 3
C. 5
D. 6

_______________ **85.** Beam test specimens should be cured at a temperature between ___.
A. 20°F to 40°F
B. 40°F to 60°F
C. 60°F to 80°F
D. none of the above

Completion

______ **1.** The maximum level of turbidity for mix water is ___ parts per million.

______ **2.** When testing fine aggregate for organic impurities, approximately 4½ oz of sand is added to the test bottle and the bottle is filled to about 7 oz with a 3% solution of ___.

______ **3.** The amount of water-reducing admixture added to a concrete mixture is usually less than ___% by weight of the cement.

______ **4.** A(n) ___ is a water-reducing admixture that provides concrete with increased workability with less mix water.

______ **5.** The amount of calcium chloride used in a concrete mixture should be less than ___% by weight of the cement.

______ **6.** A sulfate concentration of ___% may reduce concrete strength by 4%.

______ **7.** When testing fine aggregate for organic impurities and its impact on mortar strength, fine aggregate is acceptable if mortar made with it is not less than ___% of the strength of mortar made with washed sand.

______ **8.** Calcium chloride is a crystalline solid ___ admixture.

______ **9.** Tricalcium silicate and dicalcium silicate comprise approximately ___% of cement.

______ **10.** A(n) ___, also known as a high-range, water-reducing admixture, significantly reduces the amount of water required in a concrete mixture or greatly increases the slump of concrete without severely impacting setting time or air entrainment.

______ **11.** A(n) ___ is a fine particle substance that chemically reacts with calcium hydroxide.

______ **12.** A(n) ___ facilitates expansion setting and is used in nonshrink grouts.

______ **13.** Water-reducing admixtures give each of the cement particles a(n) ___ charge and increase the hydration efficiency of the cement to 80%.

______ **14.** ___ air is not intentionally incorporated into the concrete mixture.

______ **15.** A(n) ___ admixture is added to a concrete patching mixture to help the mixture adhere to the area being patched.

______ **16.** ___ is the ability of fresh concrete to flow.

______ **17.** A(n) ___ decreases air content in concrete mixtures so hardeners can be cast on a fresh slab and incorporated into the surface.

______ **18.** Maximum size aggregate should not exceed ___ the depth of slabs.

______ **19.** For adequate durability, normal weight concrete that is subjected to freeze/thaw cycles while wet should be properly air-entrained and have a water-cement ratio of approximately ___ by weight.

______ **20.** The specific gravity of cement is ___.

______________ **21.** A(n) ___ specification describes the exact proportions of all ingredients in a concrete mixture.

______________ **22.** The ___ process is a process in which concrete is mixed in the mixing drum of a truck traveling to the placement site.

______________ **23.** ___ specifications place the responsibility of proper mixture design on the concrete supplier.

______________ **24.** Most specifications allow a tolerance of ± ___% in weight or volume of admixtures.

______________ **25.** The ___ process is a process in which concrete is partially mixed at a plant and the remainder of mixing is completed in a ready-mix or agitator truck traveling to the placement site.

______________ **26.** A(n) ___ is a crack caused by rapid evaporation of moisture as it leaves the surface at a faster rate than bleeding occurs.

______________ **27.** A(n) ___ surface contains very shallow waves, similar to light ripples on the surface of a body of water.

______________ **28.** When conducting a beam test for concrete containing large aggregate, the minimum cross-sectional dimension should be greater than ___ times the maximum size of the aggregate.

______________ **29.** The ___ is a measure of the ultimate load-carrying capacity of a beam.

______________ **30.** A(n) ___ is used to measure flexural strength of concrete using a standard unreinforced beam.

______________ **31.** A(n) ___ test is used to determine the compressive strength of concrete.

______________ **32.** ___ is usually performed when the compressive strength of concrete is not known or when compression tests produce questionable results.

______________ **33.** A(n) ___ mixture is wet concrete that clings to floats and knee boards.

______________ **34.** Slump is measured to the nearest ___″.

______________ **35.** A substance applied to existing concrete that creates a bond between it and the next layer of concrete or patch is a(n) ___.

______________ **36.** A(n) ___ is a hardened concrete protrusion that extends from the surface.

______________ **37.** A(n) ___ is a concrete defect, usually in the shape of a flake, that becomes detached from a concrete surface due to a blow to the concrete, environmental exposure, pressure, or expansion within the concrete.

______________ **38.** ___, also known as chattering, generally occurs because the blade pitch of a power trowel is too great during the first finish troweling.

______________ **39.** At least ___ days of satisfactory drying conditions should be allowed before exposing a slab to de-icing chemicals.

______________ **40.** ___ is a deposit of salts on the surface emerging from substances within the concrete.

______________ **41.** To reduce the potential for spalling in cold environments, the water-cement ratio should not exceed ___.

______________ **42.** ___ is a variation in the color of the concrete surface from what is normal or desired.

______________ **43.** ___ is a concrete repair process in which zero slump (or near zero slump) concrete, mortar, or grout is forced into a confined space.

______________ **44.** The deflection of a concrete member due to sustained loads is ___.

______________ **45.** Flowing mixtures have a slump greater than ___″.

______________ **46.** When conducting a unit weight test, use a(n) ___ cu ft container for concrete mixtures containing aggregate larger than 2″.

______________ **47.** For adequate drainage, slabs are typically sloped approximately ___″ per 4′ of lineal run.

______________ **48.** ___ is a concrete surface finishing technique that results in an even and smooth texture across the surface.

______________ **49.** A(n) ___ occurs when concrete on one or both sides of a joint or crack buckles upward.

______________ **50.** The air temperature for the 24 hr period prior to sack rubbing and for 72 hr following the procedure should be approximately ___°F to ensure that moisture does not evaporate too quickly.

______________ **51.** ___ is a hand rubbing process done with a carborundum stone that raises a coat of mortar paste that fills bug holes, pitting, or light honeycomb.

______________ **52.** When grout is applied under pressure to a crack or joint, the grout pressure should be restricted to ___ psi.

______________ **53.** A slump cone should be filled in ___ equal layers.

______________ **54.** A unit weight test is used during concrete placement to determine if changes in ___ or mixture proportions have occurred.

______________ **55.** When conducting a unit weight test, use a(n) ___ cu ft container for concrete mixtures containing aggregate up to 2″ in size.

______________ **56.** The ___ method for performing an air content test is a practical method for testing all concrete but is especially useful for concrete made with lightweight and porous aggregate.

______________ **57.** Two compression test specimens are tested at ___ days.

______________ **58.** ___ is the measured maximum resistance of concrete to axial loading.

______________ **59.** When molding compression test specimens, concrete with a slump less than ___″ must be vibrated.

Short Answer

1. Why should workers placing and finishing concrete containing an accelerating admixture be notified that this material is in the mixture?

2. Describe the effect that an air-entraining admixture has on concrete.

3. List four techniques that can be used to prevent or minimize rock pockets and honeycomb.

4. List six measures that can be taken to avoid crazing and map cracking.

5. How can waviness be controlled?

____

____

____

____

____

____

Matching

Concrete Defects

______ 1. Rock pocket

______ 2. Honeycomb

______ 3. Bug hole

______ 4. Pitting

______ 5. Crazing

______ 6. Sand streak

______ 7. Popout

______ 8. Scaling

A. Disintegration and flaking of surface mortar on a hardened concrete surface

B. Shallow, conical depression in a concrete surface that remains after a small piece of concrete has broken away due to internal pressure

C. Brownish streaks that resemble long, fine tree roots

D. Network of very fine and shallow cracks that form irregular patterns in the surface of concrete

E. Small holes in a concrete surface caused by corrosion and disintegration

F. Voids in vertical wall surfaces caused by water or air trapped against the form surface

G. Void left in concrete due to mortar not effectively filling the space among coarse aggregate

H. Porous void in hardened concrete that consists primarily of coarse aggregate and open voids with little or no mortar

Concrete Defect Identification

______________ **1.** Honeycomb

______________ **2.** Cracks

______________ **3.** Popouts

______________ **4.** Bug holes

______________ **5.** Scaling

Ⓐ

Ⓑ

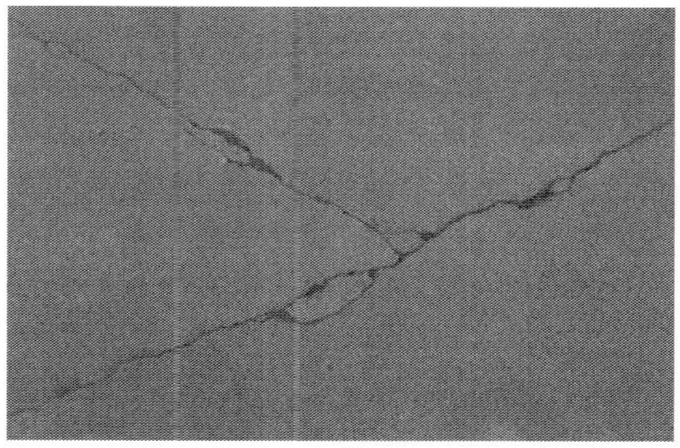

Ⓒ

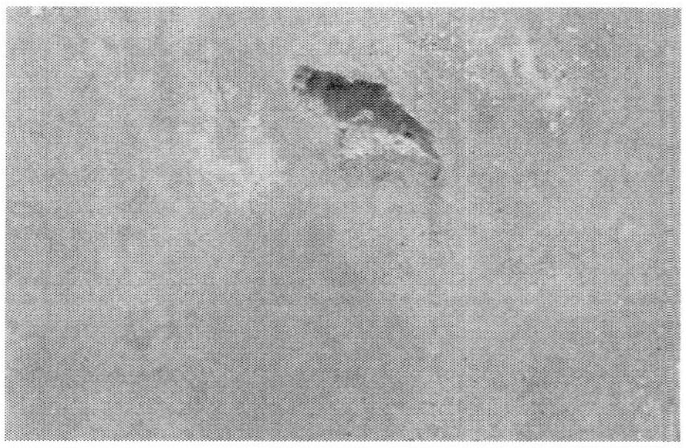

Ⓓ

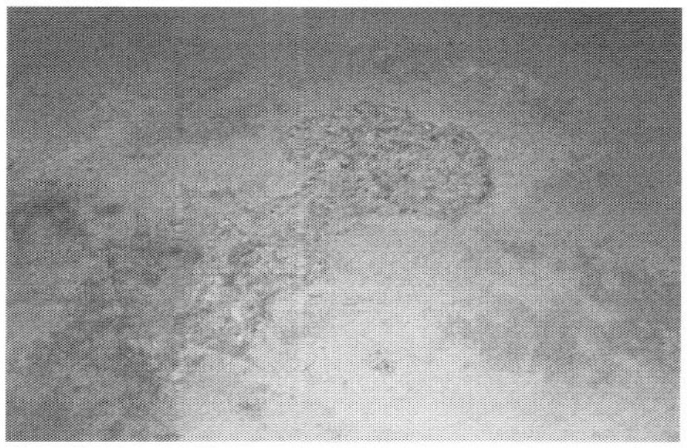

Ⓔ

Activities

Compressive Strength

Compressive strength is the measured maximum resistance of concrete to axial loading, which is typically expressed in pounds per square inch (psi). Compressive strength is calculated using the pressure gauge reading from a hydraulic press used to perform the compression test and the cross-sectional area of the compression test specimen. Compressive strength can be calculated by using the following formula:

$$f'_c = \frac{f_f}{A}$$

where

f'_c = compressive strength (in psi)
f_f = pressure gauge reading at failure (in lb)
A = cross-sectional area of specimen (in sq in.)

For example, determine the compressive strength of a concrete specimen measuring 6″ in diameter that breaks with a gauge reading of 125,000 lb.

1. Determine cross-sectional area. Cross-sectional area is determined by applying the formula:

   $A = 0.7854d^2$

   where

   A = area (in sq in.)
   0.7854 = constant
   d = diameter (in in.)

   $A = 0.7854d^2$
   $A = 0.7854 \times (6 \times 6)$
   $A = 0.7854 \times 36$
   $A = 28.27$ sq in.

2. Determine compressive strength.

   $$f'_c = \frac{f_f}{A}$$

   $$f'_c = \frac{125,000}{28.27}$$

   $$f'_c = \mathbf{4421.6\ psi}$$

Determine the compressive strength of the following compression test specimens:

________ **1.** Pressure gauge reading at failure = 145,000 lb; specimen = 6″ diameter

________ **2.** Pressure gauge reading at failure = 157,000 lb; specimen = 6″ diameter

________ **3.** Pressure gauge reading at failure = 225,000 lb; specimen = 8″ diameter

________ **4.** Pressure gauge reading at failure = 250,000 lb; specimen = 8″ diameter

________ **5.** Pressure gauge reading at failure = 445,000 lb; specimen = 12″ diameter

Projects

1. Contact a ready-mixed concrete supplier and discuss the mixtures commonly specified by customers. Determine the types of admixtures used most often and how the admixtures are added to the concrete mixtures.
2. Contact a concrete contractor to identify the types of concrete defects most commonly encountered. Discuss techniques and procedures that can be used to reduce or eliminate the defects.

Concrete Principles

Chapter 10

Concrete Construction Estimating

Name ______________________________ Date ____________________

True-False

T F **1.** Labor pricing is based on the number of hours required and the cost of labor per hour.

T F **2.** Labor, material, and equipment costs must be determined before quantity takeoff is performed.

T F **3.** Division 03 of the specifications commonly contains a listing of bid categories.

T F **4.** Dimensions on construction prints are expressed as engineering units or architectural units.

T F **5.** An estimating software system is a computer program organized like a spreadsheet that performs takeoff calculations when dimensions of structures are entered.

T F **6.** The memory method of inch-decimal foot conversion is referred to as field conversion.

T F **7.** Site plans are commonly noted with the prefix S.

T F **8.** When dividing a topographical plan into grids, the size of a grid is determined by the nature of the terrain.

T F **9.** When forming piers, formwork is only required on the inside of the pier.

T F **10.** A solid contour line on a site plan is used to represent the finished grade.

T F **11.** Excavation takeoff for site utilities is based on the linear feet of trenching required to install the desired utility.

T F **12.** The cross-section method of calculating excavated material is used when the sides of the excavation are not parallel.

T F **13.** When estimating plywood form components for concrete walls, the total area is determined by multiplying the wall height by the wall length.

T F **14.** The total length of stake material required for footing wall forms is determined by multiplying the number of stakes by the depth of the footing.

T F **15.** The number of studs for wall forms is calculated by dividing the length of the wall by the on-center distance between studs.

T F **16.** When estimating piped utilities, the calculations are commonly made in cubic feet of pipe required.

T F **17.** The formwork for elevated slabs may be calculated as the square feet of contact area of the underside of the slab.

T F **18.** Information concerning formwork is typically provided in structural drawings.

T F **19.** The volume of concrete required for columns is computed by multiplying the cross-sectional area of the column by the height of the column from floor to ceiling.

T F **20.** When estimating form materials, dimensions are rounded to the next highest foot increment before calculations are performed.

T F **21.** When using 2″ thick planks as sheathing for footings, the total perimeter of the footing must be determined.

T F **22.** Footing volume is based on footing depth, width, and length.

T F **23.** Base plates are commonly used as a base for the inner form walls of job-built or patented wall systems.

T F **24.** Welded wire reinforcement is available in 750 sq ft rolls and 32 sq ft sheets.

T F **25.** The size of the upright rebar and ties and the spacing of the rebar in both directions are indicated on the column schedule or in the detail drawings.

T F **26.** Concrete wall volume is based on wall thickness, height, and length dimensions of the wall.

T F **27.** Rebar is commonly priced as the cost per foot.

T F **28.** The volume of concrete for slabs is found by multiplying the area (in sq ft) by the slab depth (in in.).

Multiple Choice

____________________ **1.** In a prebid meeting, all parties interested in a construction project ___.

A. review the project
B. question the architect or owner
C. share information
D. all of the above

____________________ **2.** Division ___ of the CSI MasterFormat™ contains information regarding concrete.

A. 02
B. 03
C. 05
D. 06

________________ **3.** A(n) ___ is a change to the originally issued contract.
A. punch list
B. change order
C. addendum
D. none of the above

________________ **4.** The formula for calculating the volume of square or rectangular structures is ___.
A. $\frac{t \times w \times l}{27}$
B. $l \times w$
C. $\frac{0.7854d^2 \times h}{27}$
D. $0.7854d^2$

________________ **5.** A(n) ___ sheet is a grid consisting of rows and columns on a sheet of paper into which item descriptions, code numbers, quantities, and costs are entered.
A. spread
B. ledger
C. grid
D. estimating

________________ **6.** When calculating the amount of concrete required for a structure, a waste allowance of ___ should be added to each calculation to allow for spillage during placement.
A. 1% to 2%
B. 3% to 5%
C. 10% to 15%
D. 20% to 25%

________________ **7.** Full inches and decimal equivalents of an inch are divided by ___ to obtain a decimal foot conversion.
A. 6
B. 12
C. 24
D. 27

________________ **8.** Concrete volume for stairways is calculated by multiplying the number of treads by the ___ of the stairway.
A. height
B. width
C. length
D. all of the above

________________ **9.** A ___ is cast on top of another slab, jacked into position, and secured to columns at the desired elevation.
A. flat plate
B. lift slab
C. capital
D. flat slab

_______________ **10.** Labor rates include all labor costs such as ___.
A. wages
B. taxes
C. fringe benefits
D. all of the above

_______________ **11.** A contour line is a ___ or solid line on a site plan used to show elevations of existing grade and finished grade.
A. dashed
B. wavy
C. dotted
D. red

_______________ **12.** The formula for calculating the area of a circular surface is ___.
A. $\frac{t \times w \times l}{27}$
B. $l \times w$
C. $\frac{0.7854d^2 \times h}{27}$
D. $0.7854d^2$

_______________ **13.** Dimensions for footings and foundation walls are given on ___.
A. foundation plans
B. elevation drawings
C. details
D. all of the above

_______________ **14.** Form tie spacing is determined from the ___.
A. patented form manufacturer recommendations
B. elevation drawings
C. specifications
D. all of the above

_______________ **15.** The quantity takeoff for excavation is typically calculated in ___.
A. square feet
B. square yards
C. cubic feet
D. cubic yards

_______________ **16.** When using the field conversion method, each full 1″ is assumed to equal ___′.
A. 0.008
B. 0.08
C. 0.8
D. 8

__________ **17.** When calculating the area of a slab, the ___ are multiplied together.

A. length and width
B. width and thickness
C. length and thickness
D. thickness, width, and length

__________ **18.** If welded wire reinforcement must be lapped, ___ additional welded wire reinforcement is required.

A. 1% to 2%
B. 3% to 5%
C. 10% to 15%
D. 20% to 25%

__________ **19.** The formula for calculating the area of a square or rectangular surface is ___.

A. $\frac{t \times w \times l}{27}$
B. $l \times w$
C. $\frac{0.7854d^2 \times h}{27}$
D. $0.7854d^2$

__________ **20.** The three dimensions of a concrete structure, such as a beam, are ___.

A. thickness, width, and depth
B. width, length, and height
C. thickness, width, and length
D. none of the above

__________ **21.** A site plan is also referred to as a ___.

A. civil drawing
B. site survey
C. site drawing
D. all of the above

__________ **22.** Form ties are calculated as a specified number per ___ of wall area.

A. cubic foot
B. square foot
C. cubic yard
D. square yard

__________ **23.** Walers are spaced at a distance determined by ___.

A. wall width
B. rate of concrete placement
C. other structural considerations in formwork design
D. all of the above

__________ **24.** The formula for calculating the volume of a cylindrical structure is ____.

A. $\frac{t \times w \times l}{27}$

B. $l \times w$

C. $\frac{0.7854d^2 \times h}{27}$

D. $0.7854d^2$

__________ **25.** Estimators add ____ of the total amount of form materials to compensate for waste.

A. 1% to 2%

B. 3% to 5%

C. 5% to 15%

D. 20% to 25%

__________ **26.** Labor costs vary greatly depending on ____.

A. geographic location

B. craft

C. market conditions

D. all of the above

__________ **27.** Concrete volume calculations are based on a ____ measurement.

A. cubic foot

B. square foot

C. cubic yard

D. square yard

__________ **28.** Welded wire reinforcement is represented on detail drawings with ____ lines.

A. dashed

B. solid

C. dotted

D. red

__________ **29.** ____ is the process of leveling high spots and filling low spots of earth on a construction site.

A. Excavating

B. Grubbing

C. Cradling

D. Grading

__________ **30.** The ____ method is used to calculate the volume of excavated material when the sides of the excavation are irregularly shaped and are not parallel.

A. cross-section

B. parallel side

C. average end area

D. none of the above

_______________ **31.** ___ is the surface measurement within two boundaries and is expressed in square units.

A. Volume
B. Area
C. Surface volume
D. none of the above

_______________ **32.** The length of braces attached at a 45° angle to the form is determined by multiplying the height from the base of the footing to the attachment point by ___.

A. 0.41
B. 1
C. 1.41
D. 2

Completion

_______________ **1.** ___ is the computation of construction project costs.

_______________ **2.** A(n) ___ is a computer program that uses cells organized into rows and columns and performs various mathematical calculations when formulas and data are entered.

_______________ **3.** Field conversion assumes that each ⅛″ equals ___′.

_______________ **4.** ___ is the practice of reviewing contract documents to determine quantities of materials that are included in a bid.

_______________ **5.** ___ is the division of a topographical site plan of large areas to be excavated or graded into small squares or grids.

_______________ **6.** The ___ is a master list of numbers and titles for organizing information about construction requirements, products, and activities into a standard sequence.

_______________ **7.** When estimating concrete volume, the cubic footage value must be divided by ___ to obtain the volume in cubic yards.

_______________ **8.** A cost ___ is a compilation of a number of cost items from various sources combined into a common table for reference use.

_______________ **9.** The ___ is the distance across the circumference of a circular figure passing through the center.

_______________ **10.** A(n) ___ line is a dashed or solid line on a site plan used to show elevations of existing grade and finished grade.

_______________ **11.** ___ is the three-dimensional capacity of an object.

_______________ **12.** ___ is the process of removing stumps, roots, and undergrowth from an area.

_______________ **13.** The temporary supporting of existing utility lines in or around an excavated area to protect the lines is ___.

__________ **14.** The number of stakes needed for footing forms is determined by dividing the length of the wall by the recommended spacing and adding ___ stake(s).

__________ **15.** Braces are commonly attached at a(n) ___° angle.

__________ **16.** The area of a plywood panel is ___ sq ft.

__________ **17.** ___ is shown on section views with solid circles for end views and solid lines for edge views.

Short Answer

1. Why must an estimate be carefully prepared? What can occur when material quantities and costs are overestimated? What can occur when material quantities and costs are underestimated?

2. Outline the standard estimating procedure.

Matching

Spandrel Beam Formwork

______ **1.** Double waler

______ **2.** Slab sheathing

______ **3.** Shore brace

______ **4.** Walkway brace

______ **5.** Beam bottom

______ **6.** Safety railing

______ **7.** Stud

______ **8.** Integral slab edge form

______ **9.** Walkway

______ **10.** Kicker

______ **11.** Shore

______ **12.** Top of concrete floor slab

______ **13.** Joist

______ **14.** Stiffener

______ **15.** Cleat

______ **16.** Wedge

______ **17.** Beam sides

______ **18.** Ledger

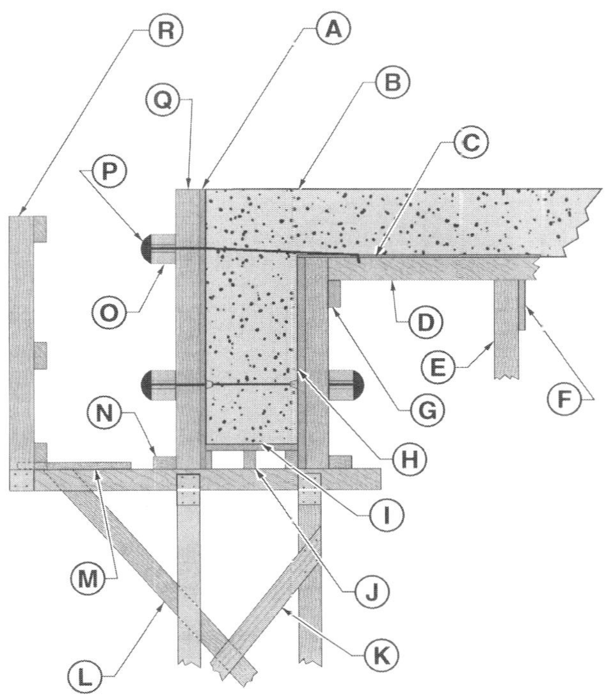

Projects

1. Obtain a copy of an estimate for a concrete structure and the corresponding print and specifications for the structure. Review the estimate, print, and specifications to determine how the estimate was performed and how accurate it is.

2. Visit various web sites, including the site for the Bureau of Reclamation and U.S. Department of Commerce, to obtain cost indexes related to concrete construction. Compare this information to cost indexes in print format and determine which sources are more up-to-date.

3. Obtain a set of prints for a multistory commercial construction project that includes concrete columns. Review the detail drawings and column schedule and identify components such as rebar splices, tie spacing, rebar arrangement, height of the columns, and similar items. Determine how the columns are tied into the various floors of the structure.